Occupational Safety, Health and Fire Index

OCCUPATIONAL SAFETY AND HEALTH

A Series of Reference Books and Textbooks
on Occupational Hazards • Safety • Health •
Fire Protection • Security • and Industrial Hygiene

Series Editor

ALAN L. KLING

Loss Prevention Consultant
Jamesburg, New Jersey

1. Occupational Safety, Health and Fire Index *David E. Miller*

Additional volumes in preparation

Occupational Safety, Health and Fire Index

A Source Guide to Voluntary and Obligatory Regulations, Codes, Standards and Publications

DAVID E. MILLER

Manager, Safety and Fire Protection

Security and Safety Department

Mobil Oil Corporation

New York, New York

MARCEL DEKKER, INC. New York and Basel

To Alice and Todd

MARCEL DEKKER, INC.
270 Madison Avenue, New York, New York 10016

LIBRARY OF CONGRESS CATALOG CARD NUMBER: 76-12285

ISBN: 0-8247-6462-5

Current printing (last digit):
10 9 8 7 6 5 4 3 2 1

PRINTED IN THE UNITED STATES OF AMERICA

FOREWORD

The hazard control problem always has been more complex than was generally recognized, but in recent years (most notably since the late 1960s) many developments happened which began to change its profile.

Under the prodding of a new advocacy, which emerged largely from outside the professional safety and health field, a greater demand occurred for safety effectiveness. The public's tolerance of injurious events apparently had worn thin. This is manifested by the quality of laws directly and indirectly related to safety that have been enacted by Congress in the past ten years.

The consequent shifting of priorities has resulted in a pervasive body of safety requirements which literally threatens the engulfment of hazard control specialists as well as management. The administrators of commerce, industry and government in particular may be hard pressed from now on to abide by the ethical, technical and legal requirements imposed on their operations by the new demands. Certainly complete, accessible references which expedite use of the many and growing requirements are needed if the necessary hazard control decisions are to be made with precision.

A fundamental step in the implementation of safety has always been the issuance of standard requirements for controlling hazards. Doubtless it will continue to be a necessary aid, since standards provide basic guidelines and tend to foster uniform technical performance. Also they should simplify the achievement tasks. However, the number and variety of agencies furnishing legal, regulatory, and technological criteria for safety and health accomplishment presently seem to compound what was already a difficult assignment.

It has been quite clear for some time that safety and health cannot be achieved where a "business as usual" climate prevails. Such attributes are the antithesis of complacency but their employment may be thwarted by the difficulties encountered when searching for applicable control requirements. Indeed, if all that is available to guide safety and health fulfillment were used fully, the number of injuries and environmental illnesses might be close to the irreducible minimum. However, the network of regulations which should support safety and health achievement may instead be a discouragement unless the prescriptives can be readily obtained.

It was to expedite the researching of hazard control criteria for its staff that prompted the Mobil Oil Corporation Security and Safety Department to prepare an Index of the nation's safety, health and fire standards. Experience with using the Index suggested improvements and its utility resulted in the decision to make it available to specialists and corporations everywhere.

Anyone who has had the responsibility for determining which requirements correctly governed a hazardous exposure and has had to sift through the seeming morass of references, will appreciate having a codified, easy to use, and regularly up-dated guide to the applicable control requirements. The provision of such a reference is the purpose of this publication.

Users of the Index will find that material such as the U.S. Coast Guard regulations is not included. This is not an oversight, but reflects the plan to cover the occupational safety, health, and fire standards which *at this time* have the most universal application. In future supplements and editions the compilers will expand the coverage as they work at the unrelenting responsibility for maintaining the Index's freshness and value.

While acknowledging the merit of this Index of requirements and the further values that regular supplements will provide, the implications for hazard control achievement through the application of standards deserves some explanation. So many agencies rely on standards and regulations as a means of reducing injuries that there may be an inate danger in such emphases. It derives from the limitations that are inherent in most written requirements and the resulting unwarranted trust that is often placed in their hazard control capabilities.

Except for one class of standards that is discussed below, most standards cannot be regarded as having an absolute ability to furnish the wanted controls. Constant vigilance and inquisitiveness about the relative effectiveness of hazard control procedures, in addition to the meticulous application of the standard requirements, are necessary if the wanted results are to be obtained.

Generally the employment of standards has had notable success in maintaining uniformity and desired performance criteria. Moreover, when a standard can be addressed specifically to hazards that essentially are controllable through technological means (and are not ordinarily alterable by an operator) they in effect provide absolute control if used. Such a standard is one of two types and may be labelled a *standard for form and structure.* It is used, for example, in the design and testing of pressure vessels, elevators, and aircraft, or where the materials of construction, integrity of design, and other physical necessities are identifiable, sharply proscribed, and within the bounds of technological ability to eliminate the established failure points. This application is the essence of prevention and its effectiveness has led to a desire to replicate its success. The trouble is that oftentimes control is not simply a function of the safeness of the structure or form under consideration.

When the failure points are intrinsically physical, known, and correctible, it is possible to write requirements which practically assure fulfillment of the hazard elimination objective. However, where safeness depends wholly or in part on the quality of operator performance, the protocol may be as important as the physical standard. Such a requirement is a *standard for procedure.*

Standards for form and structure have a superior effectiveness advantage over the standards for procedure. Besides correcting practically irreversibly a system's inherent danger points, they are also more apt to be used. There are several reasons for this. One is that standards for form and structure are concerned mostly with serious hazards. Sensible practice, therefore, dictates compliance with them for ethical and other practical reasons. Even if not mandated through legislation, observance is essential to the avoidance of public disfavor and lawsuits. On the other hand, standards for procedure even when specified by law, are subject to operator diligence. Otherwise their prescriptives are not apt to be effective. Common experience indicates how inconstant human attention to directives may be. Consequently it is desirable to maintain a differentiating posture towards the relative effectiveness of safety and health standards. Where clearly the application of a standard for form and structure does not occur, the observance of the suggested caveat is advised strongly.

This publication's objective plainly is to foster the knowing application of safety, health, and fire standards. If it is used perseveringly and wisely it can be expected to make the hazard control objective more attainable, since ultimately the effective control of hazards is the result of intensive study and the pursuit of perfection.

John V. Grimaldi

The Center for Safety
New York University

PREFACE

With the passage of the U.S. Occupational Safety and Health Act of 1970 and with similar legislation being considered overseas, Mobil Oil Corporation began to see a need for a quick reference index, or listing, of safety, health and fire codes, standards, guides and publications that could be used in determining and developing an appropriate level of safety for the workplace.

We at Mobil, therefore, have pooled our knowledge and experience to develop what we believe is a useful tool, both for the safety and health professional and for line management which, ultimately, has the responsibility for accident and injury prevention activities on a day-to-day basis. This publication is the result. It is an alphabetical list of representative applicable codes, standards, regulations, etc., in the field of occupational safety and health, and is the only publication of its kind to bring this information together for quick and easy reference.

While directed primarily to those in industry who are involved in safety and health activities, the Index should also be helpful to engineering, legal, educational, governmental, and labor groups throughout the country in developing and conducting programs on Occupational Safety and Health. Faculty and students in colleges and universities will also find the Index to be a valuable research tool.

While every effort has been made to ensure that this Index is up to date, it must be remembered that new information is being developed continually, so the listings in this publication ought not to be regarded as all-inclusive. However, as new information becomes available, it will be published annually in cumulative supplements. Those planned for the future will incorporate additional material from such groups as the Department of Transportation, U.S. Coast Guard, the Nuclear Regulatory Commission, and others. A pocket is provided inside the back cover to hold future supplements.

This Index is designed to provide information believed to be accurate, but it should be understood that this publication does not constitute legal advice. If such advice is required, a lawyer should be consulted. While every attempt is made to avoid errors, it should also be understood that all liability for loss resulting from any errors in the Index is hereby disclaimed.

Finally, this publication is dedicated to the safety and health professional and to all others who are responsible for accident prevention activities. We are of the opinion that the Index will enhance the ability to develop programs that will reduce injuries, fatalities and disease in the United States and other countries throughout the world.

Acknowledgments

I wish to thank the many people who have played a part in the production of this Index. Space limitations preclude naming all names, but the invaluable contributions of three Mobil people and one independent consultant must be noted:

John J. McKenna, Manager, Security and Safety Department, Mobil Oil Corporation—for initial recognition of the importance of the Index and for continuing advice and encouragement.

Arthur D. Puebla of Mobil Oil Corporation and Mildred S. Trask of Mobil Research and Development Corporation—for their technical assistance and untiring efforts in compiling the Index and putting it into final form for publication.

Alan L. Kling, an independent loss prevention consultant—for his guidance and insight as editorial consultant.

To these and all others who had a hand in compiling the Index, I am deeply indebted.

David E. Miller

USING THE INDEX

Major items are arranged alphabetically, as are the sub-category items within each major entry point. The only difference is that major terms have been organized so that chemical prefixes are ignored but sub-category items are in computer-sorted order. This means that a reference such as 2-ACETYLAMINOFLUORENE as a major item is found with the A's, but as a sub-term falls at the end of the alphabetical listing. (For an example of this see "AIR CONTAMINANTS.")

Seven specific reference sources are listed at the top of each page, and below each is the number of a cited standard or the abbreviated title of a publication. There is also one column with the heading "MISC" which gives the abbreviated names of organizations and their publications. (Full names can be found in the list of abbreviations and reference sources on page xi.)

If more than one publication is cited for any particular index term, each reference is printed under the appropriate column on the same line as the index term. However, if there are multiple entries in the *same column,* then the publications are printed underneath each other and the parenthetical expression "(MULTIPLE REFERENCES)" is printed beneath the index term.

To make full use of this Index it is, of course, necessary to have ready access to the cited references. Complete addresses from which these may be obtained are therefore given in the list of reference sources. Most organizations will also provide a free catalog of their publications.

An important feature of this publication is that it serves not only as a reference source but also as a *workbook.* Throughout the Index, a blank column is provided so that the user can enter applicable company information, or references from other sources or government bodies, and thereby relate this in-house data to official regulations and standards. Typical information would include company safety policy and procedure manuals, job instruction sheets, engineering guides and data references, and medical and first-aid documents. Thus, the Index provides the possibility of developing a composite listing of official regulations and standards together with in-house supporting information properly related in a single overview.

ABBREVIATIONS AND REFERENCE SOURCES

AAR	Association of American Railroads 1920 L Street Washington, DC 20036
AATCC	American Association of Textile Chemists & Colorists P.O. Box 12215 Research Triangle Park, NC 27709
AIHA	American Industrial Hygiene Association 66 South Miller Road Akron, OH 44313
ANSI	American National Standards Institute 1430 Broadway New York, NY 10018
API	American Petroleum Institute 2101 L Street Washington, DC 20037
APM	*Accident Prevention Manual*
AQM	*Air Quality Monogram*
ARI	Air-Conditioning and Refrigeration Institute 1815 North Fort Myer Drive Arlington, VA 22209
ASAE	American Society of Agricultural Engineers 2950 Niles Road St. Joseph, MI 49085
ASHRAE	American Society of Heating, Refrigerating and Air-Conditioning Engineers 345 East 47th Street New York, NY 10017

ASTM	American Society for Testing & Materials 1916 Race Street Philadelphia, PA 19103
CC	*Chem-Cards*
CGA	Compressed Gas Association 500 Fifth Avenue New York, NY 10036
CIC	*Cargo Information Cards*
HG	*Hygienic Guides* (see AIHA above)
IEC	International Electrotechnical Commission 1 rue de Varembe 1211 Geneva 20 Switzerland
IEEE	Institute of Electrical and Electronics Engineers 345 East 47th Street New York, NY 10017
IES	Illuminating Engineering Society 345 East 47th Street New York, NY 10017
ISDS	*Industrial Safety Data Sheets*
ISO	International Organization for Standardization Central Secretariat 1 rue de Varembe Case Postale 56 1211 Geneva 20 Switzerland
ITE	Institute of Traffic Engineers 1815 North Fort Myer Drive Arlington, VA 22209
MCA	Manufacturing Chemists Association 1825 Connecticut Avenue Washington, DC 20009
MRR	*Medical Research Reports*
NBS	National Bureau of Standards Standards Development Services Section Washington, DC 20234

NFPA	National Fire Protection Association 470 Atlantic Avenue Boston, MA 02210
NSC	National Safety Council 425 N. Michigan Avenue Chicago, IL 60611
PSD	*Petroleum Safety Data*
RP	*Recommended Practice*
SAE	Society of Automotive Engineers Two Pennsylvania Plaza New York, NY 10001
SD	*Safety Data*
SG	*Safety Guide*
TB	*Technical Bulletins*
TC	*Tank Car*
TEG	*Transportation Emergency Guide*
TR	*Toxicological Reviews*
UL	Underwriters' Laboratories, Inc. 207 East Ohio Street Chicago, IL 60611
USDI	United States Department of Interior 18th & C Street N.W. Washington, DC 20240
USDL	United States Department of Labor Occupational Safety and Health Administration Washington, DC 20210

Occupational Safety, Health and Fire Index

APPLICABLE STANDARDS, GUIDES AND PUBLICATIONS

SUBJECT	USDL & USDI	NFPA	ANSI	API	MCA		NSC	MISC
ABRASIVES, COATED							ISDS452	
ABRASIVE BLASTING							ISDS433	
ABRASIVE BLASTING PAINT REMOVER	1915.23							
(MULTIPLE REFERENCES)	1916.23							
AIR SUPPLY & AIR COMPRESSORS	1910.94							
BLAST CLEANING ENCLOSURES	1910.94							
DUST HAZARDS	1910.94							
EQUIPMENT	1915.23							
(MULTIPLE REFERENCES)	1916.23							
EXHAUST SYSTEM	1910.94							
GUARDING	1910.243							
NOZZLES	1910.244							
PERSONAL PROTECTIVE EQUIPMENT	1915.23							
(MULTIPLE REFERENCES)	1916.23							
PROTECTIVE EQUIPMENT	1910.94							
SAFETY CODE			Z9.4					
VENTILATION			Z9.4					
ABRASIVE WHEELS & TOOLS	1915.74							
(MULTIPLE REFERENCES)	1916.74							
(MULTIPLE REFERENCES)	1917.74							
(MULTIPLE REFERENCES)	1926.303							
FLANGES	1910.215							
GENERAL REQUIREMENTS	1910.215							
GUARDING	1910.215							
(MULTIPLE REFERENCES)	1910.243							
MOUNTING	1910.215							
POWER TOOLS, HAND & PORTABLE	1910.243							
USE, CARE, PROTECTION			B7.1					
WHEEL TERMS	1910.241							
ACCESS TO EXITS	1910.37							
BARGE OR TOWBOAT	1918.23							
CARGO SPACES	1915.46							
(MULTIPLE REFERENCES)	1916.46							
(MULTIPLE REFERENCES)	1917.46							
DRAFTS	1918.21							
EXITS & FIRE EQUIPMENT	1915.51							
(MULTIPLE REFERENCES)	1916.51							
(MULTIPLE REFERENCES)	1917.51							
FLOATING DRYDOCKS	1915.45							
(MULTIPLE REFERENCES)	1916.45							
(MULTIPLE REFERENCES)	1917.45							
ILLUMINATION	1915.52							
(MULTIPLE REFERENCES)	1916.52							
(MULTIPLE REFERENCES)	1917.52							
(MULTIPLE REFERENCES)	1918.21							
STAGING	1915.41							
(MULTIPLE REFERENCES)	1916.41							
(MULTIPLE REFERENCES)	1917.41							

SUBJECT	USDL & USDI	NFPA	ANSI	API	MCA		NSC	MISC
	APPLICABLE STANDARDS, GUIDES AND PUBLICATIONS							
(CON'T)ACCESS TO EXITS								
VESSELS	1918.11							
VESSELS AFLOAT	1915.44							
(MULTIPLE REFERENCES)	1916.44							
(MULTIPLE REFERENCES)	1917.44							
WELDING IN CONFINED SPACES	1915.31							
(MULTIPLE REFERENCES)	1916.31							
(MULTIPLE REFERENCES)	1917.31							
ACCIDENT								
CRITICALITY ACCIDENT ALARM SYSTEM			N16.2					
DOSIMETRY FOR CRITICALITY ACCIDENTS			N13.3					
MOTOR VEHICLE - CLASSIFICATION MANUAL			D16.1					
MOTOR VEHICLE - RECORDING & MEASURING			D15.1					
(MULTIPLE REFERENCES)			D15.2					
OFF-THE-JOB DISABLING ACCIDENTAL INJURY			Z16.3					
ACCIDENT PREVENTION								
ABRASIVE WHEELS	1910.215							
API FIRST AID INSTRUCTOR TRAINING GUIDE				BULL-2019				
COMPRESSED AIR	1910.169							
CONTAMINANTS	1910.93							
DRIVER IMPROVEMENT COURSE				RP-2020				
FIRST AID TRAINING				PUB-2017				
FIRST AID TRAINING (FLIP CHART)				PUB-2018				
OIL REFINERY MAINTENANCE				RP-2007				
RECEIVERS	1910.169							
RESPONSIBILITIES	1926.20							
SIGN SPECIFICATIONS			Z35.1					
(MULTIPLE REFERENCES)			Z35.4					
SIGNS, TAGS	1910.145							
TAG SPECIFICATIONS			Z35.2					
ACCIDENT REPORTING								
ANALYSIS							ISDS527A	
OFF-THE-JOB							ISDS601	
OFF-THE-JOB DISABLING ACCIDENTAL INJURY			Z16.3					
UNDERWATER ACCIDENT REPORT FORM			Z86.2					
ACCREDITATION								
ACTION	1919.4							
APPLICATION	1919.3							
CRITERIA	1919.6							
DURATION	1919.5							
RECONSIDERATION	1919.9							
RENEWAL	1919.5							
REVOCATION	1919.8							
SHORE-BASED EQUIPMENT	1919.50							
(MULTIPLE REFERENCES)	1919.51							
TERMINATION	1919.7							

APPLICABLE STANDARDS, GUIDES AND PUBLICATIONS

SUBJECT	USDL & USDI	NFPA	ANSI	API	MCA		NSC	MISC
ACCREDITED PERSONS (SEE ALSO COMPETENT PERSONS)								
APPROVAL	1919.60							
CERTIFICATE ISSUANCE	1919.10							
DUTIES	1919.10							
ELIGIBILITY	1919.50							
RECORDKEEPING	1919.11							
(MULTIPLE REFERENCES)	1919.12							
TEST SUPERVISION	1919.10							
ACETALDEHYDE					CC-49			
(MULTIPLE REFERENCES)					CIC-1			
(MULTIPLE REFERENCES)					SD-43			
(MULTIPLE REFERENCES)					TEG			HG
ACETIC ACID					CIC-2			
(MULTIPLE REFERENCES)					SD-41		ISDS410A	HG
TOXIC EFFECT			Z37.39					
ACETIC ANHYDRIDE					CIC-3			
(MULTIPLE REFERENCES)					SD-15			HG
ACETONE					CC-23			
(MULTIPLE REFERENCES)					SD-87			
(MULTIPLE REFERENCES)					TEG		ISDS398	HG
TOXIC EFFECT			Z37.32					
ACETONE CYANOHYDRIN					CC-34			
(MULTIPLE REFERENCES)					CIC-4			
(MULTIPLE REFERENCES)					TEG			
ACETONITRILE					CC-64			
(MULTIPLE REFERENCES)					CIC-5			
(MULTIPLE REFERENCES)					TEG			HG
2-ACETYLAMINOFLUORENE	1910.93							
ACETYLENE					SD-7		ISDS494	HG
FEDERAL SUPPLY CONTRACTS	50-204.66							
ACETYLENE CYLINDERS	1910.102							
CHARGING PLANT GAS SYSTEM		51A						
EFFECTIVE DATES	1910.114							
FILLING SYSTEMS	1910.102							
GENERATORS	1910.102							
INSPECTION	1910.166							
PIPING SYSTEMS	1910.102							
STANDARDS SOURCES	1910.115							

OCCUPATIONAL SAFETY, HEALTH & FIRE INDEX

APPLICABLE STANDARDS, GUIDES AND PUBLICATIONS

SUBJECT	USDL & USDI	NFPA	ANSI	API	MCA		NSC	MISC
ACETYLENE GENERATORS	1910.252							
APPROVAL	1910.252							
LOCATION	1910.252							
MAINTENANCE	1910.252							
MARKING	1910.252							
OPERATION	1910.252							
PORTABLE	1910.252							
PRESSURE LIMITS	1910.252							
RATINGS	1910.252							
STATIONARY	1910.252							
ACID CARBOYS	1910.262							
ACID PLANT							ISDS210	
ACOUSTICS								
MEASUREMENTS - FREQUENCIES & BAND NUMBERS			S1.6					
ACROLEIN					CC-76			
(MULTIPLE REFERENCES)					SD-85			
(MULTIPLE REFERENCES)					TEG		ISDS436A	HG
ACRYLONITRILE					CC-15			
(MULTIPLE REFERENCES)					CIC-6			
(MULTIPLE REFERENCES)					SD-31			
(MULTIPLE REFERENCES)					TEG			HG
ADIPIC ACID							ISDS438	
ADIPONITRILE					CIC-7			
AEC LICENSES	1910.96							
AERIAL LIFTS	1926.556							
(MULTIPLE REFERENCES)	1926.952							
(MULTIPLE REFERENCES)	1926.955							
BOOM PLATFORMS	1926.556							
BURSTING SAFETY FACTOR	1926.556							
ELECTRICAL TESTS	1926.556							
FIELD MODIFICATIONS	1926.556							
LADDER TRUCKS	1926.556							
REQUIREMENTS	1926.556							
TOWER TRUCKS	1926.556							
TRAMWAYS - OPEN PIT MINE	55.10							
TRAMWAYS - SAND, GRAVEL, CRUSHED STONE OPERATION	56.10							
TRAMWAYS - UNDERGROUND MINE	57.10							
WELDING STANDARDS	1926.556							

SUBJECT	USDL & USDI	NFPA	ANSI	API	MCA		NSC	MISC
	APPLICABLE STANDARDS, GUIDES AND PUBLICATIONS							
AGRICULTURAL EQUIPMENT								
IDENTIFICATION - OPERATOR CONTROLS			B114.2					ASAE S304.2
(MULTIPLE REFERENCES)								SAE J389
SAFETY STANDARD			B114.3					ASAE S318.3
(MULTIPLE REFERENCES)								SAE J208
AGRICULTURAL OPERATIONS	1910.267							
AIR								
COMMODITY SPECIFICATION			Z86.1					CGA G-7
COMPRESSED, CLEANING	1910.242							
FLOW TEST	75.152							
QUALITY	1910.134							
RECEIVERS	1910.169							
SUPPLY	1910.94							
(MULTIPLE REFERENCES)	1910.134							
AIR COMPRESSOR								
CLEANING							ISDS379A	
AIR CONDITIONING								
FIRE CODE		90A						
RESIDENTIAL TYPE		90B						
SAFETY STANDARD			C33.14					UL 484
AIR CONTAMINANTS								
AIR QUALITY CRITERIA & HEALTH				AQM 69-2				
ALPHA-NAPHTHYLAMINE	1910.93							
AMBIENT AIR QUALITY STANDARDS FOR PARTICULATES				AQM 69-1				
ASBESTOS	1910.93							
BENZIDINE	1910.93							
BETA-NAPHTHYLAMINE	1910.93							
BIS CHLOROMETHYL ETHER	1910.93							
COAL TAR PITCH	1910.93							
COLOR CODES	1910.134							
ETHYLENEIMINE	1910.93							
INHALATION HAZARDS	71.200							
METHYL CHLOROMETHYL ETHER	1910.93							
N-NITROSODIUMETHYLAMINE	1910.93							
PHOTOCHEMICAL SMOG				AQM 69-6				
SAMPLING	71.201							
VINYL CHLORIDE	1910.93							
2-ACETYLAMINOFLUORENE	1910.93							
3,3-DICHLOROBENZIDINE & ITS SALTS	1910.93							
4-AMINODIPHENYL	1910.93							
4-DIMETHYLAMINOAZOBENZENE	1910.93							
4-NITROBIPHENYL	1910.93							
4,4'-METHYLENE BIS(2-CHLOROANILINE)	1910.93							

APPLICABLE STANDARDS, GUIDES AND PUBLICATIONS

SUBJECT	USDL & USDI	NFPA	ANSI	API	MCA		NSC	MISC
AIR CONTROLLING EQUIPMENT								
POWER PRESS	1910.217							
AIR DRILLING								
SAFETY				RP-2011				
AIR FILTER								
HIGH EFFICIENCY - SAFETY STANDARD			B132.1					UL 586
AIR LIFT HAMMER								
FORGING	1910.218							
AIR POWERED HAND TOOLS							ISDS392	
AIR QUALITY								
ATMOSPHERIC FLUORINE & VEGETATION				AQM 69-3				
CARBON DIOXIDE				AQM 69-9				
LEAD				AQM 69-11				
OPEN PIT MINE	55.5							
OXIDANTS & HEALTH EFFECTS				AQM 69-5				
PARTICULATES				AQM 69-1				
PARTICULATES & HEALTH EFFECTS				AQM 69-2				
SAND, GRAVEL, CRUSHED STONE OPERATION	56.5							
UNDERGROUND MINE	57.5							
AIR RECEIVERS								
CLEANING	1910.169						ISDS379A	
DRAINS	1910.169							
GENERAL	1910.169							
INSTALLATION	1910.169							
PRESSURE GAGES	1910.169							
TRAPS	1910.169							
VALVES	1910.169							
AIR SUPPORTED STRUCTURE								
SAFETY STANDARD		102	Z20.3					
AIRCRAFT								
CABIN CLEANING		410F						
CLEANING		410D						
ELECTRICAL MAINTENANCE		410A						
EQUIPMENT FOR PRESSURIZING PNEUMATIC SYSTEMS							ISDS477	
FIRE DEPARTMENT - CRASH FIRES		406M						
FIRE EXTINGUISHERS, HAND		408	Z112.3					
FIRE INVESTIGATORS MANUAL		422M						
FIRE PROTECTION - INTERIOR		421						
FOAM FIRE FIGHTING VEHICLE		412						
FUEL SERVICING		407	Z119.1					
FUEL SYSTEM MAINTENANCE		410C	Z119.2					
FUELING RAMP DRAINAGE		415						

SUBJECT	USDL & USDI	NFPA	ANSI	API	MCA		NSC	MISC
	APPLICABLE STANDARDS, GUIDES AND PUBLICATIONS							
(CON'T) AIRCRAFT								
HANGARS		409	Z214.1					
HELIPORT CONSTRUCTION & PROTECTION - ROOF-TOP		418						
INTERIOR FIRE PROTECTION		421						
LOADING WALKWAY		417						
OXYGEN MAINTENANCE		410B	Z227.1					
PAINTING & PAINT REMOVAL		410D						
RAMP FIRE HAZARDS			Z215.1					
RESCUE & FIRE FIGHTING - STD OPERATING PROCEDURE		402						
RESCUE & FIRE FIGHTING SERVICES		403	Z213.1					
RESCUE & FIRE FIGHTING VEHICLES		414	B128.1					
TERMINAL BUILDING		416	Z246.1					
WELDING IN HANGAR		410E						
AIRHOSES	1910.243							
AIRLESS SPRAY PAINTING							ISDS548	
AIRPORT								
TERMINAL BUILDING		416	Z246.1					
AISLES & PASSAGEWAYS	1910.22							
MATERIALS HANDLING & STORAGE	1910.176							
ALARM SYSTEMS & COMMUNICATIONS								
ALARM VALVES	1910.159							
AUDIBLE	1926.602							
AUXILIARY SIGNALING SYSTEM		72B						
CENTRAL STATION SIGNALING SYSTEM		71						
COMMUNICATIONS	75.1600							
CRITICALITY ACCIDENT			N16.2					
EXPLOSION PREVENTION SYSTEMS		69						
FIRE	1910.37							
(MULTIPLE REFERENCES)	1926.150							
FIRE ALARM SYSTEM - MUNICIPAL		73						
FIRE WARNING SYSTEM - HOUSEHOLD		74						
LOCAL PROTECTION SIGNALING SYSTEMS		72A						
MILLS	1910.216							
PROPRIETARY SIGNALING SYSTEM		72D						
REMOTE STATION SIGNALING SYSTEM		72C						
RUBBER & PLASTICS	1910.216							
SPRINKLERS	1910.159							
ALLIGATOR SHEARS							ISDS213	
ALLOY STEEL CHAINS								
OVERHEAD LIFTING							ISDS478A	

SUBJECT	USDL & USDI	NFPA	ANSI	API	MCA		NSC	MISC
	APPLICABLE STANDARDS, GUIDES AND PUBLICATIONS							
ALLYL ALCOHOL					CC-35			
(MULTIPLE REFERENCES)					CIC-8			
(MULTIPLE REFERENCES)					TEG			HG
ALLYL CHLORIDE					CC-63			
(MULTIPLE REFERENCES)					CIC-9			
(MULTIPLE REFERENCES)					SD-99			
(MULTIPLE REFERENCES)					TEG			HG
ALLYL GLYCIDYL ETHER								HG
ALUMINUM								HG
DUST EXPLOSION		65	Z12.19					
ALUMINUM CHLORIDE					SD-62			
ALUMINUM CHLORIDE, ANHYDROUS							ISDS478A	
ALUMINUM CYLINDER								
INSPECTION	1910.166							
ALUMINUM OXIDE								HG
ALUMINUM POWDER								
DUST EXPLOSION		651	Z12.11					
4-AMINODIPHENYL	1910.93							
AMINOETHYLETHANOLAMINE					CIC-10			
AMMONIA								
ANHYDROUS, STORAGE & HANDLING	1910.111							
MOUNTED ON FARM VEHICLES	1910.111							
NITRATE, STORAGE	1910.109							
PORTABLE CONTAINERS	1910.111							
TANKS, TRANSPORTING	1910.111							
AMMONIA (ANHYDROUS)					CC-44			
(MULTIPLE REFERENCES)					CIC-11			
(MULTIPLE REFERENCES)					SD-8			
(MULTIPLE REFERENCES)					TEG		ISDS478A	HG
LIQUID TRANSFER	1910.111							
MOTOR VEHICLE	1910.111							
STORAGE & HANDLING			K61.1					CGA G-2.1
STORAGE & HANDLING - BASIC RULES	1910.111							
STORAGE & HANDLING - GENERAL	1910.111							
STORAGE & HANDLING - PORTABLE DOT CONTAINERS	1910.111							
STORAGE & HANDLING - REFRIGERATED SYSTEMS	1910.111							
STORAGE & HANDLING - STATIONARY, NONREFRIG CNTRS	1910.111							

SUBJECT	USDL & USDI	NFPA	ANSI	API	MCA		NSC	MISC
	APPLICABLE STANDARDS, GUIDES AND PUBLICATIONS							
AMMONIA (AQUA)					CC-66			
(MULTIPLE REFERENCES)					CIC-12			
(MULTIPLE REFERENCES)					SD-13			
(MULTIPLE REFERENCES)					TEG			
AMMONIUM DICHROMATE					SD-45			
AMMONIUM NITRATE								
DISTANCES FROM HAZARDS	1910.109							
ELECTRIC WIRING	1910.109							
EXPLOSIVES, SEPARATION DISTANCE		492						
FIRE PROTECTION	1910.109							
SEPARATION WALLS	1910.109							
STORAGE	1910.109	490						
AMMUNITION								
SMALL ARMS - PRIMERS	1910.109							
SMALL ARMS - SMOKELESS PROPELLANTS	1910.109							
SMALL ARMS - STORAGE	1910.109							
AMYL ACETATE							ISDS208	HG
ANCHORING FIXED MACHINERY	1910.212							
ANESTHETICS								
INHALATION - GAS SYSTEM		56A						
ANGLE OF REPOSE	1926.651							
PILE DRIVING	1926.603							
ANILINE					CC-51			
(MULTIPLE REFERENCES)					CIC-13			
(MULTIPLE REFERENCES)					SD-17			
(MULTIPLE REFERENCES)					TEG		ISDS409A	HG
ANTIMONY & ITS COMPOUNDS							ISDS408	HG
ANTIMONY TRICHLORIDE (ANHYDROUS)					SD-66			
ARC WELDING & CUTTING	1910.252							
(MULTIPLE REFERENCES)	1926.351							
CABLES & CONNECTORS	1915.36							
(MULTIPLE REFERENCES)	1916.36							
(MULTIPLE REFERENCES)	1917.36							
GROUND RETURNS & MACHINE GROUNDING	1915.36							
(MULTIPLE REFERENCES)	1916.36							
(MULTIPLE REFERENCES)	1917.36							
MANUAL ELECTRODE HOLDERS	1915.36							
(MULTIPLE REFERENCES)	1916.36							
(MULTIPLE REFERENCES)	1917.36							
SHIELDING	1915.36							
(MULTIPLE REFERENCES)	1916.36							

APPLICABLE STANDARDS, GUIDES AND PUBLICATIONS

SUBJECT	USDL & USDI	NFPA	ANSI	API	MCA		NSC	MISC
(CON'T)ARC WELDING & CUTTING								
SHIELDING & OPERATING INSTRUCTIONS	1917.36							
ARCHIVES								
RECORDS CENTER PROTECTION		232AM						
AROMATIC PETROLEUM NAPHTHA				TR				
AROMATIC PETROLEUM OIL				TR				
ARSENIC & ITS INORGANIC COMPOUNDS				AQM 69-18			ISDS499	HG
ARSENIC TRIOXIDE					SD-60			
ARSINE								HG
ASBESTOS	1910.93							HG
CAUTION SIGNS & LABELS	1910.93							
CLEANUP	1910.93							
DUST	1910.19							
(MULTIPLE REFERENCES)	1926.55							
ENGINEERING CONTROLS	1910.93							
FIBER	1910.93							
FIBER EXPOSURE	1910.93							
HOUSEKEEPING	1910.93							
MEASUREMENT METHOD	1910.93							
MEDICAL EXAMINATIONS	1910.93							
MONITORING	1910.93							
PROTECTIVE EQUIPMENT	1910.93							
RECORDS	1910.93							
VENTILATION	1910.93							
WORK PRACTICES	1910.93							
ASPHALT							ISDS215	
HEALTH ASPECTS				TB-7103				
STORAGE/LOADING SAFETY GUIDE				PSD-2205				
ASSISTANT SECRETARY	1919.2							
ATOMIC ENERGY COMMISSION	1926.53							
LICENSES	1910.96							
AUDIBLE WARNING	1926.601							
AUDIOMETER (SEE ALSO SOUND LEVEL METER)								
BACKGROUND NOISE			S3.1					
SPECIFICATIONS			S3.6					IEC 177

APPLICABLE STANDARDS, GUIDES AND PUBLICATIONS

SUBJECT	USDL & USDI	NFPA	ANSI	API	MCA		NSC	MISC
AUTOMATIC SPRINKLER SYSTEMS	1910.37							
(MULTIPLE REFERENCES)	1910.159							
FIRE DEPARTMENT CONNECTIONS	1910.159							
GENERAL REQUIREMENTS	1910.159							
MAINTENANCE OF SPRINKLER SYSTEM	1910.159							
SIGNAL SYSTEMS	1910.37							
(MULTIPLE REFERENCES)	1910.160							
SPRINKLER ALARMS	1910.159							
SPRINKLER HEAD CLEARANCE	1910.159							
AUTOMOTIVE HOISTING EQUIPMENT							ISDS437	
CONSTRUCTION, CARE, USE			B153.1					
AUTOMOTIVE MECHANIC								
TRAINING - BUS/HEAVY TRUCK/TRUCK TRAILER			D18.2					
TRAINING - PASSENGER CAR/LIGHT TRUCK			D18.1					
AUXILIARY ELECTRICAL SYSTEM								
EMERGENCY LIGHTING							ISDS248A	

SUBJECT	USDL & USDI	NFPA	ANSI	API	MCA		NSC	MISC
	APPLICABLE STANDARDS, GUIDES AND PUBLICATIONS							
BACKRAILS								
HORSE	1915.41							
(MULTIPLE REFERENCES)	1916.41							
INDEPENDENT POLE METAL SCAFFOLDS	1915.41							
(MULTIPLE REFERENCES)	1916.41							
INDEPENDENT POLE WOOD SCAFFOLDS	1915.41							
(MULTIPLE REFERENCES)	1916.41							
PAINT FLOATS	1915.41							
(MULTIPLE REFERENCES)	1916.41							
REQUIREMENTS	1915.41							
(MULTIPLE REFERENCES)	1916.41							
SWINGING SCAFFOLDS	1915.41							
(MULTIPLE REFERENCES)	1916.41							
TRESTLE LADDERS	1915.41							
(MULTIPLE REFERENCES)	1916.41							
BAFFLE PLATES								
SPRAY BOOTHS	1910.107							
BAKERY EQUIPMENT	1910.263							
AIR CONDITIONING	1910.263							
BAG CHUTES AND LIFTS	1910.263							
BISCUIT EQUIPMENT	1910.263							
BLENDERS	1910.263							
BOLTING REELS	1910.263							
CONVEYORS	1910.263							
CRACKER EQUIPMENT	1910.263							
DAMPERS	1910.263							
DEFINITION	1910.263							
DIVIDERS	1910.263							
DOUGH BRAKES	1910.263							
DUMPBINS	1910.263							
FIRE EXTINGUISHERS	1910.263							
FLOUR ELEVATORS	1910.263							
FLOUR GATES	1910.263							
FLOUR HANDLING EQUIPMENT	1910.263							
FLUES	1910.263							
MACHINERY GUARDING	1910.263							
MISCELLANEOUS EQUIPMENT	1910.263							
MIXERS	1910.263							
MOULDERS	1910.263							
OVENS	1910.263							
PULVERIZERS	1910.263							
SAFETY STANDARD			Z50.1					
SCALES, FLOUR	1910.263							
SIFTERS	1910.263							
SLICERS	1910.263							
STORAGE BINS	1910.263							
VENTILATION	1910.263							
WRAPPERS	1910.263							

SUBJECT	USDL & USDI	NFPA	ANSI	API	MCA		NSC	MISC
	APPLICABLE STANDARDS, GUIDES AND PUBLICATIONS							
BALES, HANDLING	1918.81							
BAND SAWS & BAND RESAWS	1910.213							
BARGES								
ACCESS	1918.23							
(MULTIPLE REFERENCES)	1926.605							
DEFINITION	1918.3							
DOCKS	1918.37							
FIRST AID & LIFESAVING EQUIPMENT	1926.605							
MOBILE CRANES	1926.550							
MOORING	1918.23							
MOVING BY BOOMS	1918.84							
OPEN HATCHES	1918.35							
PILE DRIVING	1926.603							
WALKING	1918.37							
WORKING SURFACES	1926.605							
BARIUM & ITS INORGANIC COMPOUNDS				AQM 69-12				HG
BARKING DEVICES	1910.261							
BARREL								
GUARDING	1910.212							
SANDING MACHINES	1910.214							
STAVE SAW	1910.214							
BARRICADES & BARRIERS	1926.202							
ADJUSTABLE PRESS & DIE GUARDS							ISDS365	
BARRIER GUARD, FIXED	1910.211							
CRANES	1926.550							
ENERGIZED SUBSTATIONS	1926.957							
EXCAVATIONS	1926.651							
MOTOR VEHICLE EQUIPMENT	1926.600							
BATHING FACILITIES, LABOR CAMPS	1910.142							
BATTERIES & CHARGERS								
BATTERY ROOMS & CHARGING	1926.403							
STORAGE	1910.178						ISDS246	
BEAMERS & WARPERS							ISDS487A	
BEAMS, HATCH (STRONGBACKS)								
PLACEMENT	1918.31							
ROLLER	1918.43							
SECURED	1918.31							
(MULTIPLE REFERENCES)	1918.43							
STOWAGE	1918.43							

SUBJECT	USDL & USDI	NFPA	ANSI	API	MCA		NSC	MISC
	APPLICABLE STANDARDS, GUIDES AND PUBLICATIONS							
BEARINGS, POWER TRANSMISSION	1910.219							
BEDDING								
ELECTRICALLY HEATED			C33.52					UL 964
BELTS, ROPE & CHAIN DRIVES	1910.219							
BELTS, SAFETY	1926.104							
MATERIALS STORAGE	1926.250							
PILE DRIVING	1926.603							
SCAFFOLDING	1926.451							
STEEL ERECTION	1926.752							
BELTS, SHIFTERS, CLUTCHES, POLES, PERCHES & FASTENERS	1910.219							
BENCH & FLOOR STANDS GUARDING	1910.215							
BENDING, FORGING MACHINES	1910.211							
BENZENE				TR	CC-17			
(MULTIPLE REFERENCES)					CIC-14			
(MULTIPLE REFERENCES)					SD-2			
(MULTIPLE REFERENCES)					TEG		ISDS308A	HG
TOXIC EFFECT			Z37.4					
BENZIDINE	1910.93							
BENZOQUINONE								HG
BENZOYL PEROXIDE					SD-81			
BENZYL CHLORIDE					CC-54			
(MULTIPLE REFERENCES)					SD-69			
(MULTIPLE REFERENCES)					TEG			
BERYLLIUM & ITS COMPOUNDS	1910.252							HG
INDUSTRIAL SAFETY DATA SHEET							ISDS562A	
TOXIC EFFECT			Z37.29					
BETA RAY								
SEALED SOURCES							ISDS461	
BILGE TRUSS HOOP RING REMOVERS	1910.214							
BINS								
EXPLOSIVES STORAGE	1910.109							

SUBJECT	USDL & USDI	NFPA	ANSI	API	MCA		NSC	MISC
BIOLOGICAL HAZARDS SIGNS	1910.145							
BISPHENOL-A								HG
BLADES EXPOSURE	1910.212							
BLANKET								
RUBBER INSULATING	1910.137							
BLAST CLEANING ENCLOSURES	1910.94							
BLAST HOLES	1910.109							
BLASTING	1926.900							
ABRASIVE							ISDS433	
AGENTS, STORAGE	1926.904							
BLACK BLASTING POWDER	75.1300							
BLASTER QUALIFICATIONS	1926.901							
ELECTRIC	1926.906							
ELECTRIC, STRAY CURRENTS							ISDS289	
EXCAVATION WORK	1926.913							
FIRING	1926.909							
HYDROBLASTING							ISDS633	
LOADING	1926.905							
MUDCAPS	75.1300							
TUNNELS & SHAFTS	1926.800							
UNDERWATER	1926.912							
WATER, HIGH PRESSURE							ISDS633	
BLASTING AGENTS								
BULK DELIVERY & MIXING	1910.109							
FIRE CODE		495						
FIRE PROTECTION	1910.109							
FIXED LOCATION MIXING	1910.109							
MATERIALS HANDLING			A10.7					
MISCELLANEOUS PROVISIONS	1910.109							
SEPARATION DISTANCE		492						
STORAGE	1910.109							
(MULTIPLE REFERENCES)	1926.904		A10.7					
TRANSPORTATION			A10.7					
USE			A10.7					
BLEACHING								
PULP & PAPER MILLS	1910.261							
TEXTILES	1910.262							

SUBJECT	USDL & USDI	NFPA	ANSI	API	MCA		NSC	MISC
	APPLICABLE STANDARDS, GUIDES AND PUBLICATIONS							
BLOCKING, BLOCKS								
EXAMINATIONS	1919.30							
(MULTIPLE REFERENCES)	1919.32							
JACKS	1926.305							
MOTOR VEHICLE	1926.600							
PILE DRIVING EQUIPMENT	1926.603							
TESTING	1919.32							
BLOTTERS	1910.215							
BLOW TORCH								
GASOLINE		393						
BLOWER								
DUST, STOCK, VAPOR REMOVAL OR CONVEYING		91	Z33.1					
BOARD DROP HAMMERS	1910.218							
BOATSWAIN'S CHAIR (SEE SCAFFOLDS)								
BODY BELTS	1926.556							
BODY BELTS, SAFETY STRAPS & LANYARDS	1926.959							
BUCKLES	1926.959							
CUSHION PARTS	1926.959							
DEE RINGS	1926.959							
DEE RINGS, LINERS	1926.959							
HARDWARE	1926.959							
SNAP HOOKS	1926.959							
SNAP HOOKS, KEEPERS	1926.959							
STITCHING	1926.959							
TESTING	1926.959							
TOOL LOOPS	1926.959							
BOILERS	1926.29							
GAS UTILIZATION EQUIPMENT			Z83.3					
OPEN-PIT MINE	55.13							
PILE DRIVING EQUIPMENT	1926.603							
SAND, GRAVEL, CRUSHED STONE OPERATION	56.13							
UNDERGROUND MINE	57.13							
BOLTS								
BOLTING	1926.752							
EQUALIZER, STAVE & HEADING SAWS	1910.214							
BOOM	1910.180							
CRANE BOOM STOPS							ISDS412	
MOVING BARGES OR CARS	1918.84							
PLACEMENT	1918.54							
PLATFORMS	1926.556							

SUBJECT	USDL & USDI	NFPA	ANSI	API	MCA		NSC	MISC
	APPLICABLE STANDARDS, GUIDES AND PUBLICATIONS							
(CON'T)BOOM								
TOPPING OR LOWERING	1918.54							
BORING MACHINES	1910.213							
BORING MILLS								
HORIZONTAL METAL							ISDS269	
VERTICAL METAL							ISDS347	
BORON HYDRIDES					SD-84		ISDS508A	
BORON TRIFLUORIDE					CC-87			
(MULTIPLE REFERENCES)					TEG			
BRACKETS, ROOFING	1910.28							
(MULTIPLE REFERENCES)	1926.451							
BRAKES								
DISC-TYPE (POWER PLATFORMS)	1910.66							
EARTHMOVING EQUIPMENT	1926.602							
EQUIPMENT	1926.600							
(MULTIPLE REFERENCES)	1926.602							
HOISTING MACHINES	1910.66							
HOLDING	1910.179							
MANUALLY FED DOUGH BRAKES	1910.263							
MOTOR VEHICLE	1926.601							
POWER PRESS - CONSTRUCTION, CARE, USE			B11.3					
PRESS	1910.217						ISDS419A	
SELF-ENERGIZING BAND-TYPE (POWER PLATFORMS)	1910.66							
SHOE-TYPE (POWER PLATFORMS)	1910.66							
TRACKS & RAILROADS	1910.178							
BRAKING DEVICES								
CARGO GEAR	1919.22							
CERTIFICATION	1919.22							
BRAZING & SOLDERING								
HAND							ISDS445A	
BRICKLAYERS' SQUARE (SEE SCAFFOLDS)								
BRIDGE PLATE	1918.24							
INDUSTRIAL TRUCKS	1910.178							
BRIDLES								
BEAM & PONTOON	1918.42							
CARGO HANDLING	1918.81							
FLUSH END PALLETS	1918.67							
GANGWAYS	1918.21							

SUBJECT	USDL & USDI	NFPA	ANSI	API	MCA		NSC	MISC
	APPLICABLE STANDARDS, GUIDES AND PUBLICATIONS							
BROMINE					CC-59			
(MULTIPLE REFERENCES)					SD-49			
(MULTIPLE REFERENCES)					TEG		ISDS313	HG
BROMOCHLOROMETHANE								HG
BROMOMETHANE					CC-38			
(MULTIPLE REFERENCES)					CIC-59			
(MULTIPLE REFERENCES)					SD-35			
(MULTIPLE REFERENCES)					TEG			HG
TOXIC EFFECT			Z37.24					
BUCKETS (TUBS, HAND-LOADED)	1918.82							
BUCKETS, CONCRETE	1926.700							
BUFFING OPERATION	1910.94							
VENTILATION CONTROL			Z43.1					
BUGGIES, CONCRETE	1926.700							
BUILDINGS								
ACCESSIBILITY BY PHYSICALLY HANDICAPPED			A117.1					
AIRPORT TERMINAL		416	Z246.1					
AREAS & HEIGHTS		206M						
BUILDING TYPES, STANDARD		220						
CONSTRUCTION		241	A10.21					
DEMOLITION		241	A10.21					
GLAZING MATERIAL			Z97.1					
MAINTENANCE POWER PLATFORM	1910.66							
RECORDS	1910.66							
SAWMILLS	1910.265							
TEMPORARY	1926.151							
BUILDING FACILITIES	1910.265							
BUILDING SERVICE EQUIPMENT	1910.177							
ELECTRIC WIRING	1910.177							
BULK CARGO								
EXAMINATIONS	1919.81							
MOVING VEHICLES	1918.73							
BULK CARRIERS, GREAT LAKES	1918.35							
BULK PLANTS								
CHEMICAL PLANTS	1910.106							
COMBUSTIBLE & FLAMMABLE LIQUIDS	1910.106							
DRAINAGE	1910.106							
ELECTRIC WIRING	1910.106							

SUBJECT	APPLICABLE STANDARDS, GUIDES AND PUBLICATIONS							
	USDL & USDI	NFPA	ANSI	API	MCA		NSC	MISC
(CON'T)BULK PLANTS								
FIRE PROTECTION	1910.106							
LOADING	1910.106							
WHARVES	1910.106							
BULLDOZERS							ISDS256A	
BULLFLOATS	1926.700							
BULLING								
DEFINITION	1918.3							
LEADS	1918.84							
BUMPERS, OVERHEAD GANTRY CRANES	1910.179							
BURNS								
CHEMICAL							ISDS523	
1,3-BUTADIENE								HG
BUTADIENE								
(MULTIPLE REFERENCES)				TR	CC-81			
(MULTIPLE REFERENCES)					CIC-15			
(MULTIPLE REFERENCES)					SD-55			
					TEG			
BUTANOL (SEE BUTYL ALCOHOL)								
2-BUTANONE								
(MULTIPLE REFERENCES)					CC-22			
(MULTIPLE REFERENCES)					SD-83			
					TEG			HG
2-BUTOXYETHANOL								HG
BUTYL ACETATE								
(MULTIPLE REFERENCES)					CC-60			
					TEG			HG
BUTYL ACRYLATE					CIC-16			
BUTYL ALCOHOL								
(MULTIPLE REFERENCES)					CC-80			
					TEG			HG
BUTYLAMINE								HG
BUTYL CELLOSOLVE (SEE BUTOXYETHANOL)								
BUTYLLITHIUM								
(MULTIPLE REFERENCES)					CC-28			
					TEG			

APPLICABLE STANDARDS, GUIDES AND PUBLICATIONS

SUBJECT	USDL & USDI	NFPA	ANSI	API	MCA		NSC	MISC
N-BUTYLLITHIUM IN HYDROCARBON SOLVENTS					CC			
(MULTIPLE REFERENCES)					SD-91			
(MULTIPLE REFERENCES)					TEG			
BUTYL MERCAPTAN (SEE BUTANETHIOL)								
BUTYRALDEHYDES					CC-50			
(MULTIPLE REFERENCES)					CIC-17			
(MULTIPLE REFERENCES)					SD-78			
(MULTIPLE REFERENCES)					TEG			

SUBJECT	USDL & USDI	NFPA	ANSI	API	MCA		NSC	MISC
			APPLICABLE STANDARDS, GUIDES AND PUBLICATIONS					
CABLES & WIRES								
ARMORED			C33.9					UL 4
DIRECT BURIED UTILITY CABLE							ISDS607	
EYE SPLICES	1919.24							
FLEXIBLE CABLES & CORDS	1926.402							
GAS - TUBE - SIGN			C33.23					UL 814
IGNITION			C33.23					UL 814
RUBBER INSULATED			C33.6					UL 44
SERVICE - ENTRANCE			C33.51					UL 854
TRAILING CABLES	1926.402							
TRAILING CABLES - FIRE RESISTANT	75.600							
TRENCHING CABLES	1926.956							
WELDING CABLES & CONNECTORS	1926.351							
CABLEWAYS			B20.1					
CABS, OVERHEAD GANTRY CRANES	1910.179							
CADMIUM				AQM 69-16			ISDS312	HG
TOXIC EFFECT			Z37.5					
WELDING	1910.252							
CAGES, FIXED LADDERS	1910.27							
CAISSONS	1926.801							
CALCIUM CARBIDE	1910.252				SD-23			
CALCIUM OXIDE (SEE LIME)								
CALENDERS	1910.216							
DEFINITION	1910.262							
PROTECTED BY LOCATION	1910.216							
RUBBER & PLASTICS			B28.1					
SAFETY CONTROLS	1910.216							
CAMPHOR OIL					CIC-18			
CANOPY, PROTECTIVE	1910.266							
CAPACITORS							ISDS404A	
CAPACITY								
EXIT	1910.37							
OCCUPANT LOAD, EGRESS	1910.37							
WIDTH, EGRESS MEANS	1910.37							

SUBJECT	APPLICABLE STANDARDS, GUIDES AND PUBLICATIONS: USDL & USDI	NFPA	ANSI	API	MCA	NSC	MISC
CAR PLATE (SEE BRIDGE PLATE)							
CARBOLIC ACID (SEE PHENOL)							
CARBOLIC OIL					CIC-19		
CARBON, ACTIVATED (SEE ACTIVATED CHARCOAL)							
CARBON BISULFIDE (CARBON DISULFIDE)						ISDS341	
CARBON DIOXIDE							HG
EXTINGUISHING SYSTEM		12	A54.1				
CARBON DIOXIDE (DRY ICE)						ISDS397	
CARBON DIOXIDE EXTINGUISHING SYSTEMS	1910.161						
GENERAL REQUIREMENTS	1910.161						
INSPECTION & MAINTENANCE	1910.161						
CARBON DISULFIDE					CIC-20		
(MULTIPLE REFERENCES)					SD-12	ISDS341	HG
TOXIC EFFECT			Z37.3				
CARBON MONOXIDE	1918.93						
(MULTIPLE REFERENCES)	1918.102					ISDS415A	HG
AIR QUALITY STANDARDS				AQM 69-9			
TOXIC EFFECT			Z37.1				
CARBON TETRACHLORIDE	1910.179						
(MULTIPLE REFERENCES)	1926.150				CIC-21		
(MULTIPLE REFERENCES)					SD-3	ISDS331	HG
TOXIC EFFECT			Z37.17				
CARBOYS, ACID	1910.262						
CARGO & PORTABLE TANK STORAGE	1910.168						
CARGO GEAR							
BRAKING DEVICES	1919.22						
CHAINS	1919.25						
DAMAGED COMPONENTS	1919.20						
DEFINITION	1919.2						
DERRICK ATTACHMENT	1919.23						
GRACE PERIODS	1919.18						
HEAT TREATMENT	1919.16						
(MULTIPLE REFERENCES)	1919.17						
POWER SOURCES	1919.22						
ROPE, WIRE	1919.24						
SAFE WORKING LOADS	1919.13						
(MULTIPLE REFERENCES)	1919.21						

SUBJECT	USDL & USDI	NFPA	ANSI	API	MCA		NSC	MISC
	APPLICABLE STANDARDS, GUIDES AND PUBLICATIONS							
(CON'T)CARGO GEAR								
TESTING	1919.14							
(MULTIPLE REFERENCES)	1919.15							
WELDING	1919.19							
CARGO HANDLING (FLAMMABLE)	1910.106							
CARGO HOOKS	1918.52							
(MULTIPLE REFERENCES)	1926.551							
CARPENTERS' BRACKET (SEE SCAFFOLDS)								
CASE HOOKS	1918.66							
(MULTIPLE REFERENCES)	1918.81							
CATHODE-RAY TUBE							ISDS240A	
CAULKING (EXTERIOR)	1910.66							
CAUSTICS	1910.262							
CAUSTIC LIQUOR ROOM							ISDS214	
CAUSTIC POTASH (POTASSIUM HYDROXIDE)					CC-32			
(MULTIPLE REFERENCES)					CIC-22			
(MULTIPLE REFERENCES)					SD-10			
(MULTIPLE REFERENCES)					TEG			
TANK CAR UNLOADING					TC-3			
CAUSTIC SODA (SODIUM HYDROXIDE)					CC-33			
(MULTIPLE REFERENCES)					CIC-23			
(MULTIPLE REFERENCES)					SD-9			
(MULTIPLE REFERENCES)					TEG		ISDS373	
TANK CAR UNLOADING					TC-3			
CAUTION SIGNS & LABELS	1910.145							
ASBESTOS	1910.93							
FLUORIDES	1910.252							
RADIATION	1910.96							
WELDING	1910.252							
CELLULOSE NITRATE								
FIRE CODE		40						
CENTRIFUGES							ISDS591	

APPLICABLE STANDARDS, GUIDES AND PUBLICATIONS

SUBJECT	USDL & USDI	NFPA	ANSI	API	MCA		NSC	MISC
CERTIFICATIONS								
BOILERS	1926.29							
PRESSURE VESSELS	1926.29							
SHORE-BASED MATERIALS HANDLING DEVICES	1919.70							
CHAIN	1910.265							
ALLOY STEEL	1926.251							
ANNEALED	1918.64							
CHAINS & CHAIN SLINGS	1915.62							
(MULTIPLE REFERENCES)	1916.62							
(MULTIPLE REFERENCES)	1917.62							
FALLS & PULL-LIFTS	1910.263							
(MULTIPLE REFERENCES)	1915.64							
(MULTIPLE REFERENCES)	1916.64							
(MULTIPLE REFERENCES)	1917.64							
INSPECTION	1918.64							
KNOTS	1918.64							
LIMITATIONS	1919.25							
OVERHEAD LIFTING - ALLOY STEEL							ISDS478A	
PROOF TESTS	1918.64							
REMOVAL FROM SERVICE	1918.64							
REPAIRS	1918.64							
SAFE WORKING LOADS	1915.62							
(MULTIPLE REFERENCES)	1916.62							
(MULTIPLE REFERENCES)	1917.68							
(MULTIPLE REFERENCES)	1918.64							
SAWS	1910.266							
SHORTENED	1918.64							
CHAIN GUARDING	1910.219							
CHAIN SAW								
POWER, PORTABLE							ISDS320	
CHANGE ROOMS (PERSONNEL)	1910.141							
CHARGE INITIATION	1910.109							
CHARGERS & BATTERIES (SEE BATTERIES & CHARGERS)								
CHEMICAL								
BULK PLANTS	1910.106							
CHEMICAL PLANTS	1910.106							
CONTAINER & PORTABLE TANK STORAGE	1910.106							
DIP TANKS	1910.108							
DISTILLERIES	1910.106							
FLAMMABLES OR COMBUSTIBLES, STORAGE	1910.106							
(MULTIPLE REFERENCES)	1910.107							
INDUSTRIAL PLANTS	1910.106							
PIPING, VALVES & FITTINGS	1910.106							

SUBJECT	USDL & USDI	NFPA	ANSI	API	MCA		NSC	MISC
	APPLICABLE STANDARDS, GUIDES AND PUBLICATIONS							
(CON'T) CHEMICAL								
PROCESSING PLANTS	1910.106							
REFINERIES	1910.106							
SERVICE STATIONS	1910.106							
STORAGE TANK	1910.106							
TOILETS	1910.143							
CHEMICAL BURNS							ISDS523	
CHEMICAL PAINT REMOVERS	1915.22							
(MULTIPLE REFERENCES)	1916.22							
CHEMICAL PROCESSES								
SAFETY					SG-14			
CHEMICAL SAFETY REFERENCES							ISDS486A	
CHICKEN LADDERS	1910.28							
CHIMNEY								
FIRE CODE		211	A52.1					
REMOVAL	1926.854							
CHIPPERS & HOGS							ISDS602	
CHLORATES							ISDS371	
CHLORINATED HYDROCARBONS	1910.252							
CHLORINE					CC-53			
(MULTIPLE REFERENCES)					CIC-24			
(MULTIPLE REFERENCES)					SD-80			
(MULTIPLE REFERENCES)					TEG		ISDS207	
TOXIC EFFECT			Z37.25					
CHLORINE DIOXIDE							ISDS525	HG
CHLORINE TRIFLUORIDE					CC-2			
(MULTIPLE REFERENCES)					TEG			
(MULTIPLE REFERENCES)					TEG			
CHLOROBENZENE					CIC-25			HG
CHLORODIPHENYLS								HG
CHLOROETHANE					CC-24			
(MULTIPLE REFERENCES)					CIC-40			
(MULTIPLE REFERENCES)					SD-50			
(MULTIPLE REFERENCES)					TEG			HG

APPLICABLE STANDARDS, GUIDES AND PUBLICATIONS

SUBJECT	USDL & USDI	NFPA	ANSI	API	MCA		NSC	MISC
BIS-CHLOROETHYL ETHER	1910.93							
CHLOROFORM					CIC-26			
(MULTIPLE REFERENCES)					SD-89			HG
CHLOROHYDRIN					CIC-27			
CHLOROMETHANE					CC-67			
(MULTIPLE REFERENCES)					CIC-60			
(MULTIPLE REFERENCES)					SD-40			
(MULTIPLE REFERENCES)					TEG			HG
CHLORONAPHTHALENES								HG
CHLOROSULFONIC ACID					CC-73			
(MULTIPLE REFERENCES)					CIC-28			
(MULTIPLE REFERENCES)					SD-33			
(MULTIPLE REFERENCES)					TEG			
CHROMIC ACID & CHROMATES					SD-44			HG
TOXIC EFFECT			Z37.7					
CHROMIUM				AQM 69-15				
CHURN DRILL								
BLAST HOLE DRILLING							ISDS200	
CHUTES	1918.68							
WASTE DISPOSAL	1926.252							
CIRCUIT, LOCKOUT & TAGGING	1926.400							
CIRCUIT BREAKER								
LOW VOLTAGE - AC POWER			C37.19					
CIRCULAR SAWS								
RESAWS	1910.213							
SELF-FED	1910.213							
CLAMPS, BEAM OR FRAME	1918.84							
CLASSIFICATION								
ACCIDENT - MOTOR VEHICLE			D16.1					
EPOXY PRODUCTS BY HAZARD			K68.1					
FIRE RESISTANCE - RECORDS PROTECTION EQUIPMENT			Z134.1					
FLAMMABLE LIQUID	1910.106	321						

SUBJECT	USDL & USDI	NFPA	ANSI	API	MCA		NSC	MISC
	APPLICABLE STANDARDS, GUIDES AND PUBLICATIONS							
CLEANING								
AIR COMPRESSOR							ISDS379A	
AIR RECEIVERS	1910.169						ISDS379A	
AIRCRAFT		410D						
AIRCRAFT CABIN		410F						
BULK OXYGEN SYSTEMS	1910.104							
COMPRESSED AIR	1910.242							
CONTAINERS FOR COMBUSTIBLES							ISDS432	
ELECTRIC MOTORS							ISDS285A	
FLAMMABLE LIQUID TANK CAR				APM-13				
FLAMMABLE LIQUID TANK VEHICLE				RP-2013				
FUR STORAGE		81						
MACHINERY							ISDS285A	
PETROLEUM STORAGE TANK				RP-2015				
POWDER COATINGS	1910.107							
POWERED PLATFORMS	1910.66							
RESPIRATORS	1910.134							
SEWER PIPE							ISDS577	
SMALL TANKS - COMBUSTIBLE LIQUIDS		327						
SOLVENTS	1910.107							
SPRAY	1910.94							
SPRAY BOOTHS	1910.107							
SPRAYING OPERATIONS	1910.107							
WINDOW			A39.1					
WITH HOT WATER & STEAM							ISDS238	
CLEANING COMPOUNDS	1910.252							
COMBUSTIBILITY OF SWEEPING COMPOUNDS			Z240.1					UL 41
DEGREASING	1910.252							
MANUFACTURER'S INSTRUCTIONS	1910.252							
CLEARING, CLEARANCES								
CRANES & DERRICKS	1910.179							
(MULTIPLE REFERENCES)	1910.181							
(MULTIPLE REFERENCES)	1926.550							
(MULTIPLE REFERENCES)	1926.950							
(MULTIPLE REFERENCES)	1926.955							
FIRE PREVENTION	1926.151							
FIXED LADDERS	1910.27							
MANLIFTS	1910.68							
MATERIALS STORAGE	1926.250							
MECHANICAL HANDLING EQUIPMENT	1910.176							
SITE	1926.604							
SPRAYING DISCHARGES	1910.107							
SPRINKLER SYSTEMS	1910.159							
STAIRS	1910.24							
TEMPORARY HEATING DEVICES	1926.154							
WORKSPACE	1926.400							

APPLICABLE STANDARDS, GUIDES AND PUBLICATIONS

SUBJECT	USDL & USDI	NFPA	ANSI	API	MCA		NSC	MISC
CLIMBING EQUIPMENT	1926.951							
BODY BELTS	1926.951							
INDUSTRIAL SAFETY DATA SHEET							ISDS606	
LANYARDS	1926.951							
(MULTIPLE REFERENCES)	1926.959							
LIFELINES	1926.951							
ROPE	1926.951							
SAFETY LINES	1926.951							
STRAPS, SAFETY	1926.951							
CLIPPER								
VENEER							ISDS542	
CLIPS, WIRE	1910.179							
CLOTHING (SEE PROTECTIVE EQUIPMENT)								
CLUTCHES	1910.217							
DEFINITION	1910.180							
POWER TRANSMISSION APPARATUS	1910.219							
COAL								
DUST EXPLOSION - PROCESSING PLANT		653	Z12.7					
COAL MINES - UNDERGROUND								
ABANDONED AREAS	75.1701							
AIR FLOW TEST	75.152							
CERTIFIED PERSON	75.100							
ELECTRICAL WORK	75.153							
MAPS	75.1200							
METHANE & OXYGEN DEFICIENCY TESTS	75.150							
METHANE TEST	75.151							
OIL & GAS WELLS	75.1700							
SAFETY	75.1							
SMOKING PROHIBITION	75.1702							
TRAINING PROGRAMS	75.160							
COAMING CLEARANCES								
INTERMEDIATE DECKS	1918.41							
NOT REQUIRED	1918.41							
TRUNK HATCHES	1918.41							
WEATHER DECKS	1918.41							
COAMING ROLLERS, SECURED	1918.52							
COATINGS	1917.33							
CHEMICAL REMOVAL	1915.22							
(MULTIPLE REFERENCES)	1916.22							
FLAMMABLE LIQUID	1915.25							
(MULTIPLE REFERENCES)	1916.25							

SUBJECT	USDL & USDI	NFPA	ANSI	API	MCA		NSC	MISC
	APPLICABLE STANDARDS, GUIDES AND PUBLICATIONS							
(CON'T) COATINGS								
FLOW	1910.108							
MECHANICAL REMOVAL	1915.23							
(MULTIPLE REFERENCES)	1916.23							
ORGANIC - FLAMMABLE LIQUIDS		235						
POWDER	1910.107							
ROLL	1910.108							
TOXICITY SPECIFICATIONS			Z66.1					
WATER PIPES, VALVES, & FITTINGS							ISDS613	
WELDING	1915.33							
(MULTIPLE REFERENCES)	1916.33							
COBALT								HG
COFFERDAMS	1926.802							
COLD FORMER								
CONSTRUCTION, CARE, USE			B11.7					
COLD HEADER								
CONSTRUCTION, CARE, USE			B11.7					
COLD-ROOM								
DIESEL ENGINE TESTING							ISDS465	
GASOLINE ENGINE TESTING							ISDS465	
COLD WORK								
EQUIPMENT	1915.12							
(MULTIPLE REFERENCES)	1917.12							
FLAMMABLE & TOXIC SUBSTANCES	1915.12							
(MULTIPLE REFERENCES)	1917.12							
GROUNDING	1915.12							
(MULTIPLE REFERENCES)	1917.12							
PORTABLE LAMPS	1915.12							
(MULTIPLE REFERENCES)	1917.12							
SAFETY, FIRE	1915.12							
(MULTIPLE REFERENCES)	1917.12							
SAFETY, MEN	1915.12							
(MULTIPLE REFERENCES)	1917.12							
VENTILATION	1915.12							
(MULTIPLE REFERENCES)	1917.12							
WARNING SIGNS	1915.12							
(MULTIPLE REFERENCES)	1917.12							
COLLARS & COUPLINGS	1910.219							

SUBJECT	USDL & USDI	NFPA	ANSI	API	MCA		NSC	MISC
			APPLICABLE STANDARDS, GUIDES AND PUBLICATIONS					
COLOR CODES								
AIR CONTAMINANTS	1910.134							
EFFECTIVE DATES	1910.149							
GAS MASK CANISTERS	1910.134		K13.1					
IN INDUSTRY							ISDS219	
PHYSICAL HAZARDS	1910.144		Z53.1					ISO R408
SAFETY			Z53.1					
STANDARDS SOURCES	1910.147							
COMBUSTIBILITY								
SWEEPING COMPOUNDS			Z240.1					UL 41
COMBUSTIBLE LIQUIDS								
DEFINITION	1910.106							
DEMOLITION	1926.850							
DRY CLEANING PLANT			Z8.2					
FLAMMABLE & COMBUSTIBLE LIQUIDS	1926.152							
HOT WORK	1915.34							
(MULTIPLE REFERENCES)	1916.34							
(MULTIPLE REFERENCES)	1917.34							
PAINTING	1915.24							
(MULTIPLE REFERENCES)	1916.24							
PIPE CONNECTORS			B148.1					UL 567
PIPING	1910.106							
PRECAUTIONS	1915.25							
(MULTIPLE REFERENCES)	1916.25							
PROPERTIES		325M						
SAFEGUARDING EMPLOYEES	1915.21							
(MULTIPLE REFERENCES)	1916.21							
SOLVENTS	1915.21							
(MULTIPLE REFERENCES)	1916.21							
STORAGE	1910.106							
WELDING	1915.31							
(MULTIPLE REFERENCES)	1916.31							
COMBUSTIBLE MATERIALS & ROCK DUSTING								
ACCUMULATION OF COMBUSTIBLE MATERIALS	75.400							
CLEANUP	75.400-2							
DEFINITION	75.400-1							
DUST - EXCESSIVE AMOUNT	75.401-1							
DUST ABATEMENT	75.401							
ROCK DUSTING	75.402							
COMBUSTIBLE METALS								
FIRE PROTECTION							ISDS567	

SUBJECT	USDL & USDI	NFPA	ANSI	API	MCA		NSC	MISC
			APPLICABLE STANDARDS, GUIDES AND PUBLICATIONS					
COMBUSTIBLE VAPORS								
PROPERTIES		325M						
COMBUSTION ENGINE, STATIONARY								
GAS SYSTEM		37						
COMMERCIAL VEHICLES, LPG	1910.110							
COMMUNICABLE DISEASES REPORTING, LABOR CAMPS	1910.141							
COMMUNICATION LINES & CIRCUITS								
INSTALLATION & MAINTENANCE			C2.2					
(MULTIPLE REFERENCES)			C2.3					NBS H81
MANLIFTS	1910.66							
PROTECTORS			C33.22					UL 497
SAFETY STANDARD			C2.4					
COMMUNICATION SYSTEMS (SEE ALARM SYSTEMS & COMM)								
COMPETENT PERSONS (SEE ALSO ACCREDITED PERSONS)								
APPLICATION	1915.10							
(MULTIPLE REFERENCES)	1916.10							
(MULTIPLE REFERENCES)	1917.10							
ATMOSPHERIC TESTS	1915.11							
(MULTIPLE REFERENCES)	1917.11							
COLD WORK	1915.12							
(MULTIPLE REFERENCES)	1917.12							
CRITERIA	1915.10							
(MULTIPLE REFERENCES)	1916.10							
(MULTIPLE REFERENCES)	1917.10							
DEFINITION	1915.2							
(MULTIPLE REFERENCES)	1916.2							
(MULTIPLE REFERENCES)	1917.2							
(MULTIPLE REFERENCES)	1919.2							
DESIGNATION	1915.10							
(MULTIPLE REFERENCES)	1916.10							
(MULTIPLE REFERENCES)	1917.10							
FORM (MAR-8)	1915.10							
(MULTIPLE REFERENCES)	1916.10							
(MULTIPLE REFERENCES)	1917.10							
GEAR CERTIFICATION	1919.37							
INSPECTION & TEST	1917.10							
INTERNAL COMBUSTION ENGINES	1915.101							
(MULTIPLE REFERENCES)	1916.101							
(MULTIPLE REFERENCES)	1917.76							
LOGGING OF INSPECTIONS & TESTS	1915.10							
(MULTIPLE REFERENCES)	1916.10							
PAINTING	1915.24							
(MULTIPLE REFERENCES)	1916.24							
POWER & LIGHT CABLES	1915.25							
(MULTIPLE REFERENCES)	1916.25							

SUBJECT	USDL & USDI	NFPA	ANSI	API	MCA		NSC	MISC
	APPLICABLE STANDARDS, GUIDES AND PUBLICATIONS							
(CON'T)COMPETENT PERSONS (SEE ALSO ACCREDITED PERSONS)								
POWER MACHINERY	1917.67							
RETESTING	1915.14							
(MULTIPLE REFERENCES)	1917.14							
SHIPS GEAR	1915.67							
(MULTIPLE REFERENCES)	1916.67							
UNFIRED PRESSURE VESSELS	1915.101							
(MULTIPLE REFERENCES)	1916.101							
WELDING, CUTTING & HEATING HOLLOW METAL CONTAINR	1915.34							
(MULTIPLE REFERENCES)	1916.34							
(MULTIPLE REFERENCES)	1917.34							
WELDING, PRESERVATIVE COATINGS	1915.33							
(MULTIPLE REFERENCES)	1916.33							
(MULTIPLE REFERENCES)	1917.33							
COMPLIANCE RESPONSIBILITY	1918.2							
COMPRESSED AIR	1910.242							
CLEANING	1910.242							
FEDERAL SUPPLY CONTRACTS	50-204.8							
OPEN-PIT MINE	55.13							
PNEUMATIC POWER TOOLS	1926.302							
POWER TOOLS, HAND & PORTABLE	1910.242							
SAND, GRAVEL, CRUSHED STONE OPERATION	56.13							
UNDERGROUND MINE	57.13							
COMPRESSED GAS								
FEDERAL SUPPLY CONTRACTS	50-204.70							
COMPRESSED GAS CYLINDERS	1915.35							
(MULTIPLE REFERENCES)	1916.35							
(MULTIPLE REFERENCES)	1917.35							
CORROSION OR PITTING	1910.166							
CYLINDER VALVE			B57.1					CGA V-1
(MULTIPLE REFERENCES)								ISO R407
DEFINITION	1910.166							
DENTS	1910.166							
FEDERAL SUPPLY CONTRACTS	50-204.65							
FLOW CAPACITY (DEFINITION)	1910.167							
FRANGIBLE DISC (DEFINITION)	1910.167							
FUSIBLE PLUG (DEFINITION)	1910.167							
HIGH-LOW PRESSURE (DEFINITION)	1910.166							
IDENTIFICATION IN CYLINDERS				PSD-2206				
INSPECTION	1910.101							
INSPECTION, GENERAL REQUIREMENTS	1910.166							
LABELING PORTABLE CONTAINERS			Z48.1					CGA C-4
(MULTIPLE REFERENCES)								ISO R443
(MULTIPLE REFERENCES)								ISO R448
MARKINGS	1910.252							
PRESSURE OPENING (DEFINITION)	1910.167							

SUBJECT	USDL & USDI	NFPA	ANSI	API	MCA		NSC	MISC
	APPLICABLE STANDARDS, GUIDES AND PUBLICATIONS							
(CON'T)COMPRESSED GAS CYLINDERS								
RELIEF DEVICES	1910.167							
SAFETY RELIEF DEVICES	1910.101							
(MULTIPLE REFERENCES)	1910.167							
SET PRESSURE (DEFINITION)	1910.167							
STORAGE & HANDLING - UTILIZATION	1910.161							
WALL THICKNESS	1910.166							
YIELD TEMPERATURE (DEFINITION)	1910.167							
COMPRESSION MOLDING								
PLASTICS							ISDS632	
COMPRESSORS								
AIR COMPRESSOR CLEANING							ISDS379A	
PROCESS INDUSTRIES - SAFETY STANDARDS			B19.3					
COMPRESSOR PLANTS	1926.803							
COMPUTATION								
LOUDNESS OF NOISE			S3.4					ISO R357
CONCRETE & CONCRETE FORMS	1926.700							
BUCKETS	1926.700							
BUGGIES	1926.700							
FORMWORK, VERTICAL SHORING OF							ISDS628	
PORTLAND CEMENT PAVING							ISDS541	
SHORING, EXCAVATION & TRENCHING	1926.700							
(MULTIPLE REFERENCES)	1926.701							
TIMBER, TIMBERING	1926.701							
CONDUCTOR								
DEENERGIZED	1926.955							
WIRE	1926.954							
CONFINED SPACES								
DEFINITION	1915.2							
(MULTIPLE REFERENCES)	1916.2							
(MULTIPLE REFERENCES)	1917.2							
EMPLOYEE CHECKS	1917.54							
FIRE PREVENTION	1915.32							
(MULTIPLE REFERENCES)	1916.32							
(MULTIPLE REFERENCES)	1917.32							
FREQUENT CHECKS OF EMPLOYEE	1915.54							
(MULTIPLE REFERENCES)	1916.54							
GAS LEAKAGE	1915.32							
(MULTIPLE REFERENCES)	1916.32							
(MULTIPLE REFERENCES)	1917.32							
GENERAL SAFETY & HEALTH	1926.21							
PAINT REMOVAL	1915.23							
(MULTIPLE REFERENCES)	1916.23							

SUBJECT	USDL & USDI	NFPA	ANSI	API	MCA		NSC	MISC
	APPLICABLE STANDARDS, GUIDES AND PUBLICATIONS							
(CON'T)CONFINED SPACES								
PAINTING	1915.24							
(MULTIPLE REFERENCES)	1916.24							
WELDING	1915.31							
(MULTIPLE REFERENCES)	1916.31							
(MULTIPLE REFERENCES)	1917.31							
WELDING & CUTTING	1910.252							
(MULTIPLE REFERENCES)	1926.352							
(MULTIPLE REFERENCES)	1926.353							
WORKING	1915.54							
(MULTIPLE REFERENCES)	1916.54							
(MULTIPLE REFERENCES)	1917.54							
CONSTRUCTION								
AIR CONDITIONING - RESIDENTIAL		90B						
AIR CONDITIONING SYSTEM		90A						
AUTOMOTIVE LIFT			B153.1					
BUILDING AREAS & HEIGHTS		206M						
BUILDING CODE			A58.1					
BUILDING CONSTRUCTION & DEMOLITION		241	A10.21					
BUILDING MATERIALS - FIRE RETARDANT		703						
BUILDING TYPES, STANDARD		220						
CONCRETE			A10.9					
DIE CASTING MACHINE			B152.1					
DIP TANKS	1910.108	34						
DIVING IN CONSTRUCTION OPERATION							ISDS555	
FIRE DOORS & WINDOWS		80						
FIRE TESTS		251	A2.1					ASTM E119-69
HEAT VENTING GUIDE		204						
HEATING EQUIPMENT CLEARANCE		89M						
HOISTS							ISDS511A	
LADDER FOR MINES			M12.1					
LIFE SAFETY CODE		101	A9.1					
MASONRY WORK			A10.9					
OPERATION, EGRESS	1910.36							
PACKAGING EQUIPMENT			B155.1					
POWER PRESS, MECHANICAL	1910.217							
ROOF COVERINGS		203M						
SAFETY NET							ISDS608	
SMOKE VENTING GUIDE		204						
SPACE HEATING			A10.10					
SPRAY FINISHING			Z9.3					
STAIRS FOR MINES			M12.1					
VENTILATING SYSTEM		90A						
WARM AIR HEATING - RESIDENTIAL		90B						
WATER COOLING TOWERS		214						

SUBJECT	USDL & USDI	NFPA	ANSI	API	MCA		NSC	MISC
	APPLICABLE STANDARDS, GUIDES AND PUBLICATIONS							
CONTAINERIZED CARGO								
CAPACITY	1918.85							
DEFINITION	1918.85							
HANDLING	1918.85							
INSPECTION	1918.85							
PERMANENT MARKINGS	1918.85							
CONTAINERS	1915.102							
(MULTIPLE REFERENCES)	1916.102							
ANHYDROUS AMMONIA	1910.111							
CLEANING CONTAINERS FOR COMBUSTIBLES							ISDS432	
HYDROGEN	1910.103							
LPG	1910.110							
(MULTIPLE REFERENCES)	1926.153							
METAL SAFETY CAN			Z218.1					UL 30
METAL WASTE CAN			Z221.1					UL 32
OXYGEN, FUEL & GAS	1910.252							
PORTABLE TANK STORAGE	1910.106							
SAFETY RELIEF DEVICES	1910.110							
WELDING & CUTTING	1910.252							
(MULTIPLE REFERENCES)	1926.352							
(MULTIPLE REFERENCES)	50-204.72							
CONTAMINANTS	1910.93							
AIRBORNE, TLV	1926.55							
BENZIDINE	1910.93							
CONTROL SYSTEMS								
ELECTRICAL			C39.5					
GAS EQUIPMENT			Z226.1					UL 372
INTRINSICALLY SAFE CONTROL EQUIPMENT		493						
OIL-FIRED APPLIANCES			Z226.1					UL 372
CONVEYORS	1926.555							
BELT (FOR BULK MATERIAL) - EQUIPMENT							ISDS569A	
BELT (FOR BULK MATERIAL) - OPERATION							ISDS570A	
CONTROLS - POWERED	1918.69							
EXPLOSION-PROOF - POWERED	1918.69							
GUARDING - POWERED	1918.69							
POWER CABLES	1918.69							
ROLLER							ISDS528	
SAFETY CODE			B20.1					
COOKING EQUIPMENT								
COMMERCIAL - VENTILATION		96						

APPLICABLE STANDARDS, GUIDES AND PUBLICATIONS

SUBJECT	USDL & USDI	NFPA	ANSI	API	MCA		NSC	MISC
COPPER				AQM 69-21				
COPPER NAPHTHENATE				TR				
CORD REELS								
SAFETY STANDARD			C33.18					UL 355
CORROSION PROTECTION								
PIPING, VALVES & FITTINGS	1910.106							
STORAGE TANK	1910.106							
UNDERGROUND TANK	1910.106							
CORRUGATORS							ISDS356A	
COUNTERBALANCES	1910.217							
COUNTERWEIGHTS, CRANES, CRAWLERS, TRUCK	1910.180							
COUPLING & COLLARS	1910.219							
COVERS & GUARDRAILS	1910.22							
CRANES	1926.550							
(MULTIPLE REFERENCES)	1926.952							
(MULTIPLE REFERENCES)	1926.955							
AFLOAT, GUARDING SWING RADIUS	1918.55							
ASHORE, GUARDING SWING RADIUS	1918.74							
BOOM STOPS							ISDS412	
CABS, OVERHEAD GANTRY CRANES	1910.179							
CERTIFICATION, AFLOAT	1918.12							
CERTIFICATION, ASHORE	1918.13							
CERTIFICATION, CONTAINER	1918.14							
CLEARANCE	1910.179							
(MULTIPLE REFERENCES)	1910.181							
COUNTERWEIGHTS	1918.74							
CRAWLER			B30.5					
CRAWLER, LOCOMOTIVE & TRUCK CRANES	1910.180							
DEFINITION	1918.3							
(MULTIPLE REFERENCES)	1919.2							
ELECTRIC WIRING	1910.179							
EQUALIZER	1910.179							
EXAMINATIONS	1919.71							
(MULTIPLE REFERENCES)	1919.72							
(MULTIPLE REFERENCES)	1919.90							
EXTENSION LAMPS	1910.179							
FLOATING			B30.8					
GANTRY	1910.179		B30.2.0					
GENERAL REQUIREMENTS	1910.180							
HOISTING	1918.74							
HOISTS, ELECTROMAGNETS							ISDS359A	

APPLICABLE STANDARDS, GUIDES AND PUBLICATIONS

SUBJECT	USDL & USDI	NFPA	ANSI	API	MCA		NSC	MISC
(CON'T) CRANES								
HYDRAULIC, MOBILE			B30.15					
ILLUMINATION	1918.74							
INSPECTION CLASSIFICATION	1910.180							
INSPECTION, OVERHEAD GANTRY	1910.179							
LIGHTING	1910.179							
LOAD HANDLING	1910.180							
LOAD RATINGS	1910.180							
LOADING INDICATING DEVICES	1918.74							
LOCOMOTIVE			B30.5					
MAINTENANCE	1910.180							
OPERATING NEAR ELECTRIC POWERLINES	1910.180							
OVERHEAD			B30.2.0					
OVERHEAD GANTRY CRANES	1910.179							
OVERHEAD GANTRY, BRAKES	1910.179							
OVERHEAD GANTRY, DEFINITION	1910.179							
OVERHEAD GANTRY, ELECTRIC EQUIPMENT	1910.179							
OVERHEAD GANTRY, FOOTWALKS & LADDERS	1910.179							
OVERHEAD GANTRY, LOAD HANDLING	1910.180							
OVERHEAD GANTRY, MAINTENANCE	1910.179							
OVERHEAD GANTRY, ROPE INSPECTION	1910.179							
OVERHEAD GANTRY, STOPS, BUMPERS, RAIL SWEEPS	1910.179							
OVERHEAD GANTRY, TESTING	1910.179							
OVERHEAD GANTRY, WARNING DEVICES	1910.179							
OVERHEAD HOISTS			B30.16					
OVERHEAD WIRES	1910.180							
OVERLOADING PROHIBITED	1918.74							
PENDANT-OPERATED & CONSOLE-CONTROLLED							ISDS558A	
PILLAR			B30.4					
PORTAL			B30.4					
PULP, PAPER & PAPERBOARD MILLS	1910.261							
RADIUS INDICATORS	1918.74							
RATING CHARTS	1918.74							
RECORDS	1910.179							
REFUELING	1910.180							
REQUIREMENTS	1910.180							
RESISTORS	1910.179							
ROPE INSPECTION	1910.180							
RUNNING ROPES	1910.179							
(MULTIPLE REFERENCES)	1910.180							
SAFE WORKING LOADS	1919.75							
(MULTIPLE REFERENCES)	1919.76							
(MULTIPLE REFERENCES)	1919.77							
SIDE STRESS ON BOOM	1918.74							
SWITCHES	1910.179							
TESTING	1910.179							
(MULTIPLE REFERENCES)	1910.180							
TOWER			B30.4				ISDS630	
TRUCK			B30.5					
UNDERHUNG			B30.11					

APPLICABLE STANDARDS, GUIDES AND PUBLICATIONS

SUBJECT	USDL & USDI	NFPA	ANSI	API	MCA		NSC	MISC
(CON'T)CRANES								
UNIT PROOF TESTS	1919.71							
VISIBLE DEFECTS	1918.72							
CRAWLER, LOCOMOTIVE & TRUCK CRANES								
DEFINITION	1910.180							
GENERAL REQUIREMENTS	1910.180							
INSPECTION CLASSIFICATION	1910.180							
LOAD HANDLING	1910.180							
LOAD RATINGS	1910.180							
MAINTENANCE PROCEDURES	1910.180							
MATERIALS HANDLING & STORAGE	1910.180							
OPERATING NEAR ELECTRIC POWERLINES	1910.180							
RECORDS	1910.180							
REQUIREMENTS	1910.180							
ROPE INSPECTION	1910.180							
TESTING	1910.180							
CRAWLING BOARDS	1910.28							
(MULTIPLE REFERENCES)	1926.451							
CRESOL					CC-77			
(MULTIPLE REFERENCES)					CIC-29			
(MULTIPLE REFERENCES)					SD-48			
(MULTIPLE REFERENCES)					TEG			HG
CRESYLIC ACID (SEE CRESOL)								
CROSSCUT SAW, HAND-FED	1910.213							
CROTONALDEHYDE					CIC-30			
CROZING MACHINES	1910.214							
CRUDE OIL								
PIPELINE REPAIR GUIDE				PSD-2200				
CUMENE								HG
CUPS, SANITARY DRINKING	1918.94							
CUP WHEELS	1910.243							
FLARING-CUP, TYPE 11	1910.241							
STRAIGHT-CUP, TYPE 1	1910.241							
STRAIGHT-CUP, TYPE 6	1910.241							

APPLICABLE STANDARDS, GUIDES AND PUBLICATIONS

SUBJECT	USDL & USDI	NFPA	ANSI	API	MCA		NSC	MISC
CUPROUS OXIDE (SEE COPPER OXIDE)								
CURING, SPRAY FINISH	1910.107							
CUTOFF, SWING SAWS	1910.213							
CUTOFFS, FIRE	1926.150							
CUTTERS								
POWER GUILLOTINE							ISDS298	
CUTTING								
GAS SYSTEM		51B						
OXYGEN-FUEL GAS SYSTEM		51						
SAFETY STANDARD			Z49.1					
CUTTING OILS								
EMULSIONS & DRAWING COMPOUNDS							ISDS501	
CYCLOHEXANE				TR	CC-20			
(MULTIPLE REFERENCES)					SD-68			
(MULTIPLE REFERENCES)					TEG			HG
CYCLOHEXANONE								HG
CYLINDERS, COMPRESSED GAS (SEE COMPRESSED GAS CYL)								
CYLINDERS, GENERAL								
ACETYLENE	1910.102							
ALUMINUM CYLINDER INSPECTION	1910.166							
HIGH PRESSURE	1910.166							
HYDROGEN	1910.103							
LOW PRESSURE CYLINDER INSPECTION	1910.166							
LPG & GAS CONSUMING HEATERS	1910.110							
RESPIRATORS	1910.134							
WELDING & CUTTING, APPROVAL & MARKING	1910.252							
(MULTIPLE REFERENCES)	1926.350							

SUBJECT	USDL & USDI	NFPA	ANSI	API	MCA		NSC	MISC
	APPLICABLE STANDARDS, GUIDES AND PUBLICATIONS							
D.D.T.								HG
DEAD-MAN CONTROLS	1910.243							
DECKS, OILED	1918.38							
DECK LOADS								
CLEARANCE	1918.54							
HANDLINES	1918.41							
LADDERS	1918.25							
SAFE PASSAGE	1918.33							
DECK OPENINGS (SEE UNGUARDED EDGES & OPENINGS)								
DECOMPRESSION	1926.803							
CHAMBERS	1926.803							
TABLE	1926.804							
DEENERGIZED CONDUCTORS, STRINGING & REMOVING	1926.955							
DEENERGIZING LINES & EQUIPMENT	1926.950							
DISCONNECTING	1926.950							
GUARDS & BARRIERS	1926.950							
IDENTIFICATION	1926.950							
INSPECTION	1926.950							
ISOLATION	1926.950							
NOTIFICATION BY DESIGNATED EMPLOYEE	1926.950							
PROTECTIVE GROUNDS	1926.950							
SWITCHES & DISCONNECTORS	1926.950							
TAGGING	1926.950							
WORK COMPLETION	1926.950							
DEGREASER								
VAPOR							ISDS429	
DEGREASING								
CLEANING COMPOUNDS	1910.252							
LIQUIDS, METAL PARTS							ISDS537	
DEMOLITION	1926.850							
BUILDINGS		241	A10.21					
CHUTES	1926.852							
ENTRANCES	1926.850							
EXPLOSIVES	1926.860							
FLAMMABLE GASES	1926.850							
FLOORS	1926.850							
(MULTIPLE REFERENCES)	1926.860							
MECHANICAL	1926.859							
PASSAGEWAYS	1926.851							
SAFETY STANDARD			A10.6					

APPLICABLE STANDARDS, GUIDES AND PUBLICATIONS

SUBJECT	USDL & USDI	NFPA	ANSI	API	MCA		NSC	MISC
(CON'T)DEMOLITION								
STAIRWAYS	1926.851							
STEEL CONSTRUCTION	1926.858							
DERRICKS	1910.181							
(MULTIPLE REFERENCES)	1926.550							
ATTACHMENT	1919.23							
CERTIFICATION, AFLOAT	1918.12							
CERTIFICATION, ASHORE	1918.13							
CLEARANCE	1910.179							
(MULTIPLE REFERENCES)	1910.181							
DEFINITION	1910.181							
(MULTIPLE REFERENCES)	1918.3							
(MULTIPLE REFERENCES)	1919.2							
EXAMINATIONS	1919.73							
(MULTIPLE REFERENCES)	1919.74							
(MULTIPLE REFERENCES)	1919.90							
FLOATING			B30.8					
GIN POLE	1910.181							
GUARDS	1910.181							
INSPECTION	1910.181							
LOAD HANDLING	1910.181							
LOAD RATINGS	1910.181							
MAINTENANCE	1910.181							
OPERATION	1910.181							
OVERHEAD WIRES	1910.181							
RATINGS	1918.74							
RECORDS	1910.181							
REFUELING	1910.181							
REQUIREMENTS	1910.181							
ROPE INSPECTION	1910.181							
RUNNING ROPES	1910.181							
SAFE WORKING LOADS	1919.75							
(MULTIPLE REFERENCES)	1919.76							
(MULTIPLE REFERENCES)	1919.77							
SAFETY CODE			B30.6					
STATIONARY	1910.181							
TESTING	1910.18							
(MULTIPLE REFERENCES)	1910.181							
UNIT PROOF TESTS	1919.73							
VISIBLE DEFECTS	1918.74							
WINCH HEADS	1910.181							
DERRICK TRUCKS	1926.952							
DESIGN/CONSTRUCTION/FABRICATION								
METAL SCRAP PROCESSING EQUIPMENT			Z268.1					
STEEL PLATE			Z229.1					
STRUCTURAL STEEL			Z229.1					

APPLICABLE STANDARDS, GUIDES AND PUBLICATIONS

SUBJECT	USDL & USDI	NFPA	ANSI	API	MCA		NSC	MISC
DETEARING								
ELECTROSTATIC							ISDS468	
DETONATING CORDS	1926.908							
DEVULCANIZERS								
RUBBER INDUSTRY							ISDS554	
DIBORANE								HG
DIBROMOETHANE								HG
TOXIC EFFECT			Z37.31					
ORTHO-DICHLOROBENZENE					SD-54			HG
PARA-DICHLOROBENZENE								HG
TOXIC EFFECT			Z37.27					
3,3'-DICHLOROBENZIDINE & ITS SALTS	1910.93							
DICHLORODIFLUOROMETHANE					CIC-31			HG
TOXIC EFFECT			Z37.38					
DICHLOROETHANE					CC-62			
(MULTIPLE REFERENCES)					CIC-43			
(MULTIPLE REFERENCES)					SD-18			
(MULTIPLE REFERENCES)					TEG		ISDS350	HG
TOXIC EFFECT			Z37.21					
1,1-DICHLOROETHYLENE					CC-75			
(MULTIPLE REFERENCES)					CIC-81			
(MULTIPLE REFERENCES)					TEG			
DICHLOROMETHANE					CC			
(MULTIPLE REFERENCES)					SD-86			
(MULTIPLE REFERENCES)					TEG		ISDS474	HG
TOXIC EFFECT			Z37.18					
(MULTIPLE REFERENCES)			Z37.23					
DICHLOROPROPANE					CIC-32			
DICHLOROPROPENE					CIC-33			
DIES								
DESIGN, CONSTRUCTION, SETTING & FEEDING	1910.217							
FORGING HAMMER							ISDS467	
POWER PRESS							ISDS211	

SUBJECT	USDL & USDI	NFPA	ANSI	API	MCA		NSC	MISC
	APPLICABLE STANDARDS, GUIDES AND PUBLICATIONS							
DIE CASTING MACHINES							ISDS474	
CONSTRUCTION, CARE, USE			B152.1					
DIE GUARDS								
INDIVIDUAL & PRESS BARRIERS							ISDS365	
DIESEL ENGINE								
COLD-ROOM TESTING							ISDS465	
DIETHANOLAMINE					CIC-34			
DIETHYLAMINE					SD-97			HG
DIETHYLAMINE (ANHYDROUS)					CC-27			
(MULTIPLE REFERENCES)					TEG			
DIETHYLENETRIAMINE					CC-72			
(MULTIPLE REFERENCES)					CIC-35			
(MULTIPLE REFERENCES)					SD-76			
(MULTIPLE REFERENCES)					TEG			HG
DIETHYL OXIDE (SEE ETHYL ETHER)								
DIISOBUTYL KETONE								HG
DIISOPROPANOLAMINE					CIC-36			
DIKES								
BULK OXYGEN SYSTEMS	1910.104							
STORAGE TANK	1910.106							
DIMETHYLAMINE					CIC-37			
4-DIMETHYLAMINOAZOBENZENE	1910.93							
DIMETHYLBENZENE (SEE XYLENE)								
DIMETHYL ETHER					CC-61			
(MULTIPLE REFERENCES)					TEG			
DIMETHYLFORMAMIDE								HG
1,1-DIMETHYLHYDRAZINE								HG
DIMETHYL HYDRAZINE (SEE HYDRAZINE/UDMH)								

APPLICABLE STANDARDS, GUIDES AND PUBLICATIONS

SUBJECT	USDL & USDI	NFPA	ANSI	API	MCA		NSC	MISC
DIMETHYL SULFATE					CC-55			
(MULTIPLE REFERENCES)					SD-19			
(MULTIPLE REFERENCES)					TEG			
DINING FACILITIES, LABOR CAMPS	1910.142							
META-DINITROBENZENE								HG
2,4-DINITROPHENOL								HG
DINITROTOLUENES					SD-93			
DIOXANE								HG
DIP TANKS								
CONSTRUCTION	1910.108	34						
COVERS	1910.108							
DEFINITION	1910.108	34						
DISTANCES FROM HAZARDS	1910.108	34						
DRAINS	1910.108	34						
FIRE EXTINGUISHMENT	1910.108	34						
FLAMMABLE LIQUID	1910.108	34						
FOAM EXTINGUISHERS	1910.108	34						
HARDENING TANKS	1910.108							
IGNITION, ELECTRICAL & OTHER SOURCES	1910.108	34						
LIQUIDS, STORAGE & HANDLING	1910.108	34						
MAINTENANCE	1910.108	34						
OPERATION	1910.108	34						
OVERFLOW PIPES	1910.108	34						
SMOKING	1910.108	34						
SPECIAL APPLICATIONS	1910.108	34						
TEMPERING	1910.108							
VENTILATION	1910.108	34						
WATER SPRAY EXTINGUISHERS	1910.108	34						
DIPHENYL								HG
DISHWASHER								
COMMERCIAL - ELECTRIC			C33.95					UL 921
DISKS, SHIELDS & "U" GUARDS	1910.219							
DISPENSER								
FLAMMABLE LIQUID			Z216.1					UL 87
DISPOSAL (SEE WASTE DISPOSAL)								

SUBJECT	APPLICABLE STANDARDS, GUIDES AND PUBLICATIONS USDL & USDI	NFPA	ANSI	API	MCA		NSC	MISC
DISTANCES FROM HAZARDS								
AMMONIUM NITRATE	1910.109							
BULK OXYGEN SYSTEMS	1910.104							
DIP TANKS	1910.108	34						
ELECTROSTATIC SPRAYING	1910.107							
EXPLOSIVES STORAGE	1910.109							
IGNITION SOURCES, SEPARATION	1910.107							
SPRAY BOOTHS, SEPARATIONS	1910.107							
DISTILLERIES, REFINERIES & CHEMICAL PLANTS	1910.106							
WHARVES	1910.106							
DIVING OPERATIONS	1926.605							
CONSTRUCTION							ISDS555	
DOCK PLATES							ISDS318	
DOCKBOARDS	1910.30							
INDUSTRIAL TRUCKS	1910.178							
DOCUMENTATION	1919.90							
DOORS	1910.252							
FIRE TESTS		252	A2.2					ASTM E152-66
DOSIMETER								
GAMMA-RAY RADIATION			N13.5					
X-RAY RADIATION			N13.5					
DOSIMETRY								
CRITICALITY ACCIDENT			N13.3					
DOUGH BRAKES, MANUALLY FED	1910.263							
DRAFTS								
ACCESS MEANS	1918.21							
BUILDING	1918.82							
DOUBLE SLINGS	1918.81							
SAFELY SLUNG	1918.81							
VISIBILITY	1918.81							
DRAGLINES								
OPERATION OF							ISDS271	
DRAINAGE & SEWERS								
BULK PLANTS	1910.106							
FLOORS		92M						
INDUSTRIAL PLANTS	1910.106							
LABOR CAMPS	1910.142							
MATERIALS HANDLING	1910.176							

SUBJECT	USDL & USDI	NFPA	ANSI	API	MCA		NSC	MISC
	APPLICABLE STANDARDS, GUIDES AND PUBLICATIONS							
(CON'T)DRAINAGE & SEWERS								
PROCESSING PLANTS	1910.106							
SERVICE STATIONS	1910.106							
STORAGE TANK	1910.106							
DRAINS								
AIR RECEIVERS	1910.169							
DIP TANKS	1910.108	34						
DRAWING COMPOUNDS								
EMULSIONS & CUTTING OILS							ISDS501	
DREDGING								
SAFETY REQUIREMENTS			A10.15					
DRESSING ROOMS, PERSONNEL	1910.141							
DRILL								
CHURN							ISDS200	
JACK LEG							ISDS504	
REAMER, PORTABLE							ISDS497A	
DRILL PRESS								
METALWORKING							ISDS335	
DRILLING								
BLAST HOLE, CHURN DRILL							ISDS200	
DUST FROM ROCK DRILLING	70.400							
OFFSHORE DIVING SAFETY				TB-7104				
OIL/GAS WELL WITH FLAMMABLE LIQUID				PSD-2204				
OPEN-PIT MINE	55.7						ISDS573A	
SAFETY				BULL-2024				
(MULTIPLE REFERENCES)				RP-2010				
(MULTIPLE REFERENCES)				RP-2011				
SAND, GRAVEL, CRUSHED STONE OPERATION	56.7							
TUNNELS & SHAFTS	1926.800							
UNDERGROUND MINE	57.7							
DRINKING WATER	1918.94							
COOLER			C33.82					UL 399
GENERAL	71.600							
ON CONSTRUCTION JOB							ISDS399A	
DRIVES, BELT, ROPE & CHAIN	1910.219							
DRIVING								
DRIVER IMPROVEMENT COURSE				RP-2020				
TANK TRUCK DRIVER'S HANDBOOK				BULL-1609				

SUBJECT	APPLICABLE STANDARDS, GUIDES AND PUBLICATIONS							
	USDL & USDI	NFPA	ANSI	API	MCA		NSC	MISC
DRUMS	1910.212							
DRY BATH HEATER								
ELECTRIC			C33.75					UL 875
DRY CHEMICAL EXTINGUISHERS	1910.160							
DRY CHEMICAL EXTINGUISHING SYSTEM		17						
DRY CLEANING PLANT								
COMBUSTIBLE LIQUIDS		32	Z8.2					
DRYING, SPRAY FINISHES	1910.107							
DUAL COMPONENT COATINGS	1910.107							
DUCKBOARDS, CLEATED	1918.21							
DUMBWAITER			A17.1					
DUNNAGE								
CLEAR OF WORK AREA	1918.91							
NAILS	1918.91							
PLACEMENT ON DECK	1918.91							
RACKED AGAINST SWEAT BATTENS	1918.91							
SLINGING	1918.81							
STOWAGE	1918.91							
DUST	1926.55							
ASBESTOS	1910.19							
COAL MINE - RESPIRATORY PROTECTION			Z88.4					
COMBUSTIBLE MATERIALS & ROCK DUSTING	75.401							
CONDITIONS	1926.651							
FEDERAL SUPPLY CONTRACTS	50.204.50							
GENERAL	1910.94							
HEAVY CONCENTRATION	1918.93							
IN INDUSTRY							ISDS531	
PROTECTION	1918.93							
(MULTIPLE REFERENCES)	1918.102							
RESPIRABLE DUST STANDARDS	70.101							
RESPIRATORY EQUIPMENT	70.300							
ROCK DRILLING	70.400							
ROCK DUSTING	75.402							
SAMPLING PROCEDURE	70.201							
SANDBLASTING	1918.95							
STANDARDS	70.100							
(MULTIPLE REFERENCES)	71.100							

SUBJECT	USDL & USDI	NFPA	ANSI	API	MCA		NSC	MISC
	APPLICABLE STANDARDS, GUIDES AND PUBLICATIONS							
DUST EXPLOSION								
ALUMINUM POWDER MANUFACTURE		651	Z12.11					
ALUMINUM PROCESSING & FINISHING		65	Z12.19					
COAL PREPARATION PLANT		653	Z12.7					
COCOA PRODUCTION, PACKAGING, HANDLING		62	Z12.6					
CONFECTIONERY PLANT		657	Z12.18					
FLOUR & FEED MILLS		61C	Z12.3					
GRAIN ELEVATOR, TERMINAL		61B	Z12.4					
INDUSTRIAL PLANTS - FUNDAMENTAL PRINCIPLES		63	Z12.21					
MAGNESIUM POWDER PLANT			Z12.15					
MILLING AGRICULTURAL COMMODITIES		61D						
PLASTICS INDUSTRY		654	Z12.16					
PNEUMATIC CONVEYING SYSTEMS		66	Z12.22					
PREVENTION			Z12					
PULVERIZED - FUEL SYSTEM		60	Z12.1					
SPICE GRINDING PLANT		656	Z12.9					
STARCH FACTORY		61A	Z12.2					
SUGAR PRODUCTION, PACKAGING, HANDLING		62	Z12.6					
SULFUR FIRES		655	Z12.12					
WOOD FLOUR PLANT		664	Z12.20					
WOODWORKING PLANT		664	Z12.20					
DWELLINGS								
ELECTRICAL CODE - ONE- AND TWO-FAMILY		70A	C1.1					

SUBJECT	USDL & USDI	NFPA	ANSI	API	MCA		NSC	MISC
EAR								
PROTECTION	1926.101							
PROTECTORS			Z24.22					
EDGE, GUARDED	1918.32							
EDUCATION & TRAINING (SEE TRAINING & EDUCATION)								
EGRESS, MEANS OF	1910.35							
EMPLOYEE PROTECTION	1910.36							
FIRE PROTECTION	1910.37							
MAINTENANCE	1910.36							
ELECTRIC BLASTING								
STRAY CURRENTS							ISDS289	
ELECTRIC CIRCUITS								
REQUIREMENTS	1915.111							
(MULTIPLE REFERENCES)	1916.111							
ELECTRIC CONTROLS								
FOR HAZARDOUS LOCATION			C33.30					UL 698
POWER PRESS							ISDS624	
ELECTRIC CORD							ISDS385	
SAFETY STANDARD			C33.3					UL 817
ELECTRIC CRANE								
GROUNDING							ISDS287	
ELECTRIC CURRENT								
LEAKAGE			C101.1					
ELECTRIC CUTTING EQUIPMENT								
SAFETY & FIRE PROTECTION IN PETROLEUM OPERATIONS				RP-2009				
ELECTRIC EQUIPMENT	1926.403							
BATTERY ROOMS & CHARGING	1926.403							
CRANES & DERRICKS	1926.550							
DATA PROCESSING EQUIPMENT		75						
ELECTRONIC COMPUTER		75						
EQUIPMENT INSTALLATION & MAINTENANCE	1926.402							
GENERAL APPLICATION	1910.308							
GROUNDING & BONDING	1926.401							
GROUNDING - PORTABLE							ISDS299	
HAZARDOUS LOCATIONS	1926.404							
HIGH FREQUENCY		76CM						
IGNITION SOURCES, DIP TANKS	1910.108							
IGNITION SOURCES, SPRAY FINISHING	1910.107							
INTRINSICALLY SAFE CONTROL EQUIPMENT		493						

APPLICABLE STANDARDS, GUIDES AND PUBLICATIONS

SUBJECT	USDL & USDI	NFPA	ANSI	API	MCA		NSC	MISC
(CON'T)ELECTRIC EQUIPMENT								
METALWORKING MACHINE TOOLS		79						
OVERHEAD GANTRY CRANES	1910.179							
POWER OPERATED TOOLS	1926.302							
PURGED ENCLOSURES		496	C106.1					
SHOCK, PERSONNEL PROTECTION	1910.137							
TEST EQUIPMENT							ISDS496	
ELECTRIC FAN								
SAFETY STANDARD			C33.54					UL 507
ELECTRIC HAND SAW							ISDS344	
ELECTRIC HEATER								
DRY BATH HEATER			C33.75					UL 875
FOR HAZARDOUS LOCATION			C33.48					UL 823
SAFETY STANDARD			C33.11					UL 499
SPACE HEATING			C33.12					UL 573
ELECTRIC LEAKAGE								
APPLIANCES			C101.1					
ELECTRIC MOBILE EQUIPMENT								
GROUNDING							ISDS287	
ELECTRIC MOTOR								
CLEANING							ISDS285A	
FOR HAZARDOUS USE			C33.72					
(MULTIPLE REFERENCES)			C33.93					UL 674A
(MULTIPLE REFERENCES)								UL 674B
MAINTENANCE							ISDS546A	
ELECTRIC PLUG								
CONFIGURATION APPLICATION							ISDS579A	
RECEPTACLE-PLUG COMBINATION FOR HAZARDOUS USE			C33.97					UL 1010
ELECTRIC POWER LINES, CRANES-LOCO,CRAWLER,TRUCK	1910.180							
ELECTRIC RANGE								
HOUSEHOLD			C33.53					UL 858
ELECTRIC RECEPTACLE								
CONFIGURATION APPLICATION							ISDS579A	
RECEPTACLE-PLUG COMBINATION FOR HAZARDOUS USE			C33.97					UL 1010
ELECTRIC SHOCK								
TRUCK - ELECTRIC BATTERY POWERED			B56.3					UL 583

SUBJECT	USDL & USDI	NFPA	ANSI	API	MCA		NSC	MISC
	APPLICABLE STANDARDS, GUIDES AND PUBLICATIONS							
ELECTRIC SHOVEL								
GROUNDING							ISDS287	
ELECTRIC SPARK								
PETROLEUM PRODUCTS FIRE PROTECTION				RP-2003				
ELECTRIC SUBSTATION								
INDUSTRIAL							ISDS559	
ELECTRIC SUPPLY								
INSTALLATION & MAINTENANCE			C2.1					
(MULTIPLE REFERENCES)			C2.2					NBS H81
OUTLET BOX & FITTINGS FOR HAZARDOUS LOCATION			C33.27					UL 886
SAFETY STANDARD			C2.4					
UNDERGROUND - INSTALLATION & MAINTENANCE			C2.3					
ELECTRIC SWITCH								
CUTOUT - TROLLEY WIRES	75.1000							
FOR HAZARDOUS LOCATION			C33.29					UL 894
KNIFE SWITCH			C33.31					UL 363
LOCKING OUT							ISDS237A	
SNAP SWITCH			C33.40					UL 20
ELECTRIC SWITCHING PRACTICES							ISDS544	
ELECTRIC TOOLS								
STATIONARY & FIXED			C33.86					UL 987
ELECTRIC TRANSMISSION (SEE POWER TRANSMISSION)								
ELECTRIC VALVE								
FOR HAZARDOUS LOCATION			C33.83					UL 1002
ELECTRIC WELDING EQUIPMENT								
SAFETY & FIRE PROTECTION IN PETROLEUM OPERATIONS				RP-2009				
ELECTRIC WIRING								
AMMONIUM NITRATE	1910.109							
BUILDING SERVICE EQUIPMENT	1910.177							
BULK OXYGEN SYSTEMS	1910.104							
BULK PLANTS	1910.106							
CONSTRUCTION SITE							ISDS515	
CRANES	1910.179							
FLEXIBLE CORD & FIXTURE WIRE			C33.1					UL 62
HAND SPRAYING	1910.107							
IGNITION SOURCES	1910.107							
INDUSTRIAL PLANTS	1910.106							
LIQUEFIED HYDROGEN SYSTEMS	1910.103							
LPG	1910.110							
POWDER COATINGS	1910.107							

SUBJECT	USDL & USDI	NFPA	ANSI	API	MCA		NSC	MISC
	APPLICABLE STANDARDS, GUIDES AND PUBLICATIONS							
(CON'T)ELECTRIC WIRING								
PROCESSING PLANTS	1910.106							
SERVICE STATIONS	1910.106							
SPRAYING OPERATIONS	1910.107							
STORAGE ROOMS	1910.106							
ELECTRICAL STANDARD								
ELECTRICAL SWITCH LOCKOUT					SG-8			
HAZARDOUS AREA					SG-19			
NATIONAL ELECTRICAL CODE		70	C1					
NATIONAL ELECTRICAL SAFETY CODE			C2					NBS H30
ONE- AND TWO-FAMILY DWELLINGS		70A	C1.1					
RECREATIONAL VEHICLE PARKS		501D						
ELECTRICAL SYSTEMS								
AIRCRAFT ELECTRICAL MAINTENANCE		410A						
AUXILIARY EMERGENCY LIGHTING							ISDS248A	
COAL MINE - UNDERGROUND	75.153							
(MULTIPLE REFERENCES)	75.500							
DATA PROCESSING EQUIPMENT		75						
ELECTRONIC COMPUTER		75						
HIGH VOLTAGE LINES - UNDERGROUND	75.800							
HOSPITALS		76A						
INSPECTION - MODEL STATE LAW		70L						
LOW & MEDIUM VOLTAGE	75.900							
ELECTRICAL WORKERS								
PROTECTIVE EQUIP - HELMET			Z89.2					
PROTECTIVE EQUIP - RUBBER INSULATING BLANKET			J6.4					ASTM D1048-59
PROTECTIVE EQUIP - RUBBER INSULATING GLOVES			J6.6					ASTM D120-68
PROTECTIVE EQUIP - RUBBER INSULATING HOOD			J6.2					ASTM D1049-59
PROTECTIVE EQUIP - RUBBER INSULATING HOSE			J6.1					ASTM D1050-59
PROTECTIVE EQUIP - RUBBER INSULATING SLEEVES			J6.5					ASTM D1051-59
PROTECTIVE EQUIP - RUBBER MATTING			J6.7					ASTM D178-24
ELECTRICITY								
OPEN-PIT MINE	55.12							
SAFE USAGE IN HOSPITALS		76BM						
SAND, GRAVEL, CRUSHED STONE OPERATION	56.12							
STATIC							ISDS547	
UNDERGROUND MINE	57.12							
ELECTRODE								
CONDUCTOR GROUNDING	1926.954							
HOLDERS	1926.351							

SUBJECT	USDL & USDI	NFPA	ANSI	API	MCA		NSC	MISC
	APPLICABLE STANDARDS, GUIDES AND PUBLICATIONS							
ELECTROMAGNETIC RADIATION	1910.97							
MEASUREMENTS - MICROWAVE FREQUENCIES			C95.3					
ELECTROMAGNETS								
WITH CRANE HOISTS							ISDS359A	
ELECTROSTATIC								
DETEARING							ISDS468	
GROUNDING	1910.107							
PAINT SPRAYING							ISDS468	
SPRAY FINISHING APPARATUS	1910.107							
ELEVATED WORK PLATFORMS			A92.3					
VEHICLE MOUNTED	1910.67							
ELEVATOR								
INSPECTION MANUAL			A17.2					
SAFETY CODE			A17.1					
EMERGENCY								
BENZIDINE	1910.93							
CHEMICAL INDUSTRY					SG-4			
ENTRANCE/EXIT PETROLEUM STORAGE TANK				RP-2015				
EXCAVATIONS	1926.651							
LIGHTING - AUXILIARY ELECTRICAL SYSTEM							ISDS248A	
PUBLIC RELATIONS					SG-12			
SHELTERS	75.1500							
SWITCHES	1910.216							
SWITCHES, MILLS & CALENDERS	1910.216							
EMPLOYEE PROTECTION	1910.132							
EGRESS	1910.36							
EMPLOYEE RECORDS, RADIATION EXPOSURE	1910.96							
EMPLOYER RECORDS, RADIATION EXPOSURE	1910.96							
EMPLOYMENT								
MINORS	1910.217							
ENCLOSED SPACES								
COATINGS	1915.33							
(MULTIPLE REFERENCES)	1916.33							
(MULTIPLE REFERENCES)	1917.33							
CRANES & DERRICKS	1926.550							
DEFINITION	1915.2							
(MULTIPLE REFERENCES)	1916.2							
(MULTIPLE REFERENCES)	1917.2							
FIRE PREVENTION	1915.32							
(MULTIPLE REFERENCES)	1916.32							

SUBJECT	USDL & USDI	NFPA	ANSI	API	MCA		NSC	MISC
	APPLICABLE STANDARDS, GUIDES AND PUBLICATIONS							
(CON'T)ENCLOSED SPACES								
WELDING & CUTTING	1926.352							
WELDING, CUTTING & HEATING METALS, TOXICITY	1915.31							
(MULTIPLE REFERENCES)	1916.31							
(MULTIPLE REFERENCES)	1917.31							
WELDING, CUTTING & HEATING, COATINGS	1915.33							
(MULTIPLE REFERENCES)	1916.33							
(MULTIPLE REFERENCES)	1917.33							
ENCLOSURE, BLAST CLEANING ROOM	1910.94							
ENERGIZED SUBSTATIONS	1926.957							
BARRICADES & BARRIERS	1926.957							
CONTROL PANELS	1926.957							
DEENERGIZED LINES & EQUIPMENT	1926.957							
FENCES	1926.957							
LOWBOYS	1926.957							
MECHANICAL EQUIPMENT	1926.957							
STORAGE	1926.957							
WARNING SIGNS	1926.957							
WORK AUTHORIZATIONS	1926.957							
ENGINE								
FUELING (FIXED, PORTABLE, SELF-PROPELLED)				PSD-2203				
ENTRANCE/EXIT								
DEMOLITION	1926.851							
ENCLOSED SPACE					SG-10			
PETROLEUM STORAGE TANK				RP-2015				
TANKS					SG-10			
ENVIRONMENTAL CONTROLS	1926.50							
ACCIDENT PREVENTION SIGNS & TAGS	1910.145							
NONWATER CARRIAGE DISPOSAL SYSTEMS	1910.143							
SAFETY COLOR CODES	1910.144							
SANITATION	1910.141							
TEMPORARY LABOR CAMPS	1910.141							
EPICHLOROHYDRIN					CC-29			
(MULTIPLE REFERENCES)					CIC-38			
(MULTIPLE REFERENCES)					TEG			HG
EPOXY PRODUCTS								
CLASSIFICATION & LABELING BY HAZARD			K68.1					
EPOXY RESIN SYSTEMS							ISDS533A	HG

SUBJECT	APPLICABLE STANDARDS, GUIDES AND PUBLICATIONS							
	USDL & USDI	NFPA	ANSI	API	MCA		NSC	MISC
EQUIPMENT								
EARTHMOVING	1926.602							
ELECTRICAL	1926.402							
FIRE FIGHTING	1926.150							
LIFESAVING	1926.100							
MARINE	1926.605							
MECHANIZED	1926.600							
PILE DRIVING	1926.603							
RIGGING	1926.251							
ESCALATOR							ISDS516A	
SAFETY CODE			A17.1					
ETHANOL (SEE ETHYL ALCOHOL)								
ETHANOLAMINES								HG
2-ETHOXYETHANOL								HG
2-ETHOXYETHYL ACETATE								HG
ETHYL ACETATE					CC-18			
(MULTIPLE REFERENCES)					SD-51			
(MULTIPLE REFERENCES)					TEG			HG
ETHYL ACRYLATE					CC-85			
(MULTIPLE REFERENCES)					CIC-39			
(MULTIPLE REFERENCES)					SD-79			
(MULTIPLE REFERENCES)					TEG			HG
ETHYL ALCOHOL					CC-70			
(MULTIPLE REFERENCES)					TEG		ISDS391	HG
ETHYL AMYL KETONE (SEE 3-OCTANONE)								
ETHYLBENZENE								HG
ETHYL BROMIDE								HG
ETHYL BUTYL KETONE (SEE 3-HEPTANONE)								
ETHYL CHLORIDE (SEE CHLOROETHANE)								
ETHYL DICHLORIDE (SEE DICHLOROETHANE)								
ETHYLENE					SD-100			

APPLICABLE STANDARDS, GUIDES AND PUBLICATIONS

SUBJECT	USDL & USDI	NFPA	ANSI	API	MCA		NSC	MISC
ETHYLENE CHLOROHYDRIN								HG
ETHYLENE CYANOHYDRIN					CIC-41			
ETHYLENEDIAMINE					CIC-42			HG
ETHYLENE DIBROMIDE (SEE DIBROMOETHANE)								
ETHYLENE DICHLORIDE (SEE DICHLOROETHANE)								
ETHYLENE GLYCOL DINITRATE								HG
ETHYLENE GLYCOL MONOBUTYLETHER (SEE 2-BUTOXYETHANOL)								
ETHYLENE GLYCOL MONOETHYLETHER (SEE 2-ETHOXYETHANOL)								
ETHYLENE GLYCOL MONOMETHYLETHER (SEE 2-METHOXYETHANOL)								
ETHYLENEIMINE	1910.93				CIC-44			HG
ETHYLENE OXIDE					CC-36			
(MULTIPLE REFERENCES)					CIC-45			
(MULTIPLE REFERENCES)					SD-38			
(MULTIPLE REFERENCES)					TEG			HG
ETHYL ETHER					CC-16			
(MULTIPLE REFERENCES)					CIC-46			
(MULTIPLE REFERENCES)					SD-29			
(MULTIPLE REFERENCES)					TEG		ISDS396	HG
2-ETHYL-3-PROPYLACROLEIN					CIC-47			
ETHYL SILICATE								HG
EVACUATION								
RADIATION EVACUATION SIGNAL	1910.96							
RADIATION EXPOSURE SIGNAL			N2.3					
EXAMINATIONS (SEE ALSO TESTING)								
BLOCKS	1919.32							
BULK CARGO SPOUTS	1919.81							
(MULTIPLE REFERENCES)	1919.90							
BULK CARGO SUCKERS	1919.81							
(MULTIPLE REFERENCES)	1919.90							
CRANES	1919.71							
(MULTIPLE REFERENCES)	1919.72							
DERRICKS	1919.73							
(MULTIPLE REFERENCES)	1919.74							
GEAR, LOOSE	1919.31							
NONDESTRUCTIVE	1919.78							

SUBJECT	USDL & USDI	NFPA	ANSI	API	MCA		NSC	MISC
	APPLICABLE STANDARDS, GUIDES AND PUBLICATIONS							
(CON'T)EXAMINATIONS (SEE ALSO TESTING)								
REPLACEMENT	1919.30							
ROPE, WIRE	1919.79							
SUBSEQUENT TESTS	1919.30							
SUPERVISION	1919.10							
UNIT PROOF TESTS	1919.30							
EXCAVATIONS, EXCAVATING								
BLASTING	1926.913							
EQUIPMENT	1926.602							
(MULTIPLE REFERENCES)	1926.955							
(MULTIPLE REFERENCES)	1926.956							
(MULTIPLE REFERENCES)	1926.957							
GENERAL							ISDS482	
GROUND SUPPORT	1926.651							
MANHOLES	1926.956							
RUNWAYS	1926.650							
SITE CLEARING	1926.651							
TIMBER, TIMBERING	1926.651							
TRENCH							ISDS254	
TRENCHING & SHORING	1926.650							
WALKWAYS	1926.650							
EXEMPTIONS								
HEAT TREATMENT	1919.17							
SHORE-BASED EQUIPMENT	1919.50							
SHORE-BASED MATERIALS HANDLING DEVICES	1919.60							
EXHAUSTS	1926.57							
CRANES & DERRICKS	1926.550							
SPRAY FINISHING	1910.107							
TANKS, STORAGE	1910.94							
VATS & DIP TANKS	1910.108							
EXHAUST SYSTEMS (SEE ALSO VENTILATION)	1910.94							
DESIGN & OPERATION			Z9.2					
DUST, STOCK, VAPOR REMOVAL OR CONVEYING		91	Z33.1					
INSTRUMENTS FOR TESTING							ISDS431	
PERFORMANCE CHECKING							ISDS428	
VENTILATION - TEST INSTRUMENTS							ISDS431	
EXITS								
ACCESS	1910.37							
ACCESS OBSTRUCTION	1910.37							
ACCESS, EXTERIOR WAYS	1910.37							
ARRANGEMENT	1910.37							
CAPACITY	1910.37							
CHANGE IN ELEVATION	1910.37							
DISCHARGE FROM	1910.37							
EGRESS	1910.37							

APPLICABLE STANDARDS, GUIDES AND PUBLICATIONS

SUBJECT	USDL & USDI	NFPA	ANSI	API	MCA		NSC	MISC
(CON'T)EXITS								
EXIT SIGNS	1910.145							
HEADROOM	1910.37							
LIGHTING	1910.37							
MAINTENANCE	1910.37							
MARKINGS	1910.37							
PROTECTIVE ENCLOSURE	1910.37							
EXPLOSION								
DUST EXPLOSION-MILLING AGRICULTURAL COMMODITIES		61D						
HOT WORK REPAIRS TANK BOTTOM				PSD-2207				
NATURAL GAS-FIRED FURNACE		85B						
PREVENTION SYSTEMS		69						
TRUCK - ELECTRIC BATTERY POWERED			B56.3					UL 583
VENTING	1910.94	68						
EXPLOSIVES	1926.900							
AMMONIUM NITRATE - SEPARATION DISTANCE		492						
BLACK BLASTING POWDER	75.1300							
BULK STORAGE BINS	1910.109							
DEMOLITION	1926.860							
DISTANCES FROM HAZARDS	1910.109							
FIRE CODE		495	Z271.1					
FIRE PROTECTION	1910.109							
FIREWORKS LAW		494L						
LOADING	1910.109							
(MULTIPLE REFERENCES)	1926.905							
MANUFACTURE, TRANSPORTATION, STORAGE, USE			Z271.1					
MARKINGS	1910.109							
MATERIALS HANDLING			A10.7					
MOTOR VEHICLE TERMINALS		498						
MUDCAPS	75.1300							
OPEN-PIT MINE	55.6							
SAND, GRAVEL, CRUSHED STONE OPERATION	56.6							
SMOKING	1910.109							
STORAGE	1926.904		A10.7					
SURFACE TRANSPORTATION	1926.902							
TRANSPORTATION	1910.109		A10.7					
UNDERGROUND MINE	56.7							
UNDERGROUND TRANSPORTATION	1926.903							
USE			A10.7					
WHARVES	1910.109							
EXPLOSIVE ACTUATED TOOLS	1910.241		A10.3					
GUARDING	1910.243							
MARKINGS	1910.243							
POWER TOOLS, HAND & PORTABLE	1910.243							

SUBJECT	APPLICABLE STANDARDS, GUIDES AND PUBLICATIONS USDL & USDI	NFPA	ANSI	API	MCA		NSC	MISC
EXPOSURE								
AIRBORNE RADIOACTIVE MATERIALS	1910.96							
RADIATION	1910.96							
RADIATION - NUCLEAR FUEL PLANT PROTECTION			N7.2					
RADIATION - URANIUM MINES PROTECTION			N7.1					
RADIATION EVACUATION SIGNAL			N2.3					
RADIATION EXPOSURE SIGNAL			N2.3					
RADIATION RECORDS SYSTEM			N13.6					
EXTENSION LAMPS, CRANES	1910.179							
EXTRACTORS	1910.262							
EXTRUDER								
RUBBER & PLASTICS							ISDS610	
EYE PROTECTION	1910.133							
(MULTIPLE REFERENCES)	1926.102							
ABRASIVE BLASTING	1915.23							
(MULTIPLE REFERENCES)	1916.23							
ABRASIVE WHEELS	1915.74							
(MULTIPLE REFERENCES)	1916.74							
(MULTIPLE REFERENCES)	1917.74							
CLEANED & DISINFECTED	1918.101							
CORRECTIVE SPECTACLES	1918.101							
GOGGLES	1910.133							
IMPACT WORK	1915.81							
(MULTIPLE REFERENCES)	1916.81							
(MULTIPLE REFERENCES)	1917.81							
MAINTENANCE	1918.101							
OCCUPATIONAL & EDUCATIONAL			Z87.1					
PAINTING	1915.24							
(MULTIPLE REFERENCES)	1916.24							
POWDER ACTUATED FASTENING TOOLS	1915.75							
(MULTIPLE REFERENCES)	1916.75							
PRECAUTIONS	1915.81							
(MULTIPLE REFERENCES)	1916.81							
(MULTIPLE REFERENCES)	1917.81							
RADIANT ENERGY	1915.81							
(MULTIPLE REFERENCES)	1916.81							
(MULTIPLE REFERENCES)	1917.81							
REQUIRED	1918.101							
SPECIFICATIONS	1918.101							
SPECTRAL-TRANSMISSION PROPS OF PLASTICS			Z2					
WASH FOUNTAINS	1910.151							
WELDING	1910.252							
(MULTIPLE REFERENCES)	1915.31							
(MULTIPLE REFERENCES)	1916.31							
(MULTIPLE REFERENCES)	1917.31							

APPLICABLE STANDARDS, GUIDES AND PUBLICATIONS

SUBJECT	USDL & USDI	NFPA	ANSI	API	MCA		NSC	MISC
FABRICATION								
PLASTIC FOAM					SG-5			
STEEL PLATE			Z229.1					
STRUCTURAL STEEL			Z229.1					
FABRICS								
FLAME-RETARDANT TREATMENT							ISDS517A	
FACE PROTECTION	1910.133							
(MULTIPLE REFERENCES)	1926.102							
OCCUPATIONAL & EDUCATIONAL			Z87.1					
WELDING	1910.252							
FALLS (RUNNERS)								
CHAFING	1918.54							
DRUM WINDING	1918.52							
EYES	1918.52							
SECURED	1918.52							
WINCH DRUM TURNS	1918.52							
FALLS ON FLOOR							ISDS495	
FANS	1910.263							
FEDERAL SUPPLY CONTRACTS	50-204.1							
ACETYLENE	50-204.66							
COMPRESSED GAS CYLINDER INSPECTION	50-204.65							
COMPRESSED GASES	50-204.70							
GASES, VAPORS, FUMES, DUSTS, MISTS	50-204.50							
GENERAL SAFETY & HEALTH STANDARDS	50-204.2							
HYDROGEN	50-204.68							
MACHINERY GUARDING	50-204.5							
MATERIALS HANDLING & STORAGE	50-204.3							
MEDICAL SERVICES & FIRST AID	50-204.6							
NITROUS OXIDE	50-204.69							
NOISE EXPOSURE	50-204.10							
OXYGEN	50-204.67							
PERSONAL PROTECTIVE EQUIPMENT	50-204.7							
RADIATION	50-204.20							
RADIATION EXPOSURE	50-204.21							
RADIATION INCIDENT NOTIFICATION	50-204.30							
RADIATION PERSONNEL POSTING	50-204.27							
RADIATION PRECAUTIONS & MONITORING	50-204.23							
RADIATION RECORDS	50-204.32							
RADIATION SIGNS, LABELS, SIGNALS	50-204.24							
RADIATION WASTE DISPOSAL	50-204.29							
RADIOACTIVE MATERIALS STORAGE	50-204.28							
SAFETY	50-204.1							
SAFETY RELIEF DEVICES	50-204.71							
TOOLS & EQUIPMENT	50-204.4							

SUBJECT	APPLICABLE STANDARDS, GUIDES AND PUBLICATIONS							
	USDL & USDI	NFPA	ANSI	API	MCA		NSC	MISC
(CON'T)FEDERAL SUPPLY CONTRACTS								
USE OF COMPRESSED AIR	50-204.8							
WELDING & CUTTING CONTAINERS FOR COMBUSTIBLES	50-204.72							
FEEDING, FACILITIES, TEMPORARY LABOR CAMPS	1910.142							
FEET								
SAFETY SHOES			Z41.1					
FENDERS, EARTHMOVING EQUIPMENT	1926.602							
FIBER ROPES	1910.265							
FILLER METALS	1910.252							
FILTERS								
LENSES	1910.252							
RESPIRATORS	1910.134							
FIRE	1926.551							
FIRE ALARM & DETECTION SYSTEMS	1926.150							
BOXES	1910.163							
MAINTENANCE	1910.163							
SIGNAL SYSTEMS	1910.163							
SIGNAL SYSTEMS, GENERAL REQUIREMENTS	1910.163							
SIGNAL SYSTEMS, MAINTENANCE	1910.163							
FIRE BRIGADES							ISDS588	
FIRE CLASSES	1910.156							
FIRE EXTINGUISHERS (SEE FIRE PROTECTION EQUIPMENT)								
FIRE HAZARDS CLASSIFICATIONS	1910.156							
FIRE PREVENTION	1917.32							
(MULTIPLE REFERENCES)	1926.24							
(MULTIPLE REFERENCES)	1926.150							
FLAMMABLE LIQUID, STORAGE	1915.102							
(MULTIPLE REFERENCES)	1916.102							
FLAMMABLE LIQUID, USE	1915.25							
(MULTIPLE REFERENCES)	1916.25							
OIL & GREASE HAZARDS	1926.350							
OPEN-PIT MINE	55.4							
PRESERVATIVE COATINGS	1926.354							
SAND, GRAVEL, CRUSHED STONE OPERATION	56.4							
UNDERGROUND MINE	57.4							
WELDING	1915.32							
(MULTIPLE REFERENCES)	1916.32							

SUBJECT	USDL & USDI	NFPA	ANSI	API	MCA		NSC	MISC
	APPLICABLE STANDARDS, GUIDES AND PUBLICATIONS							
(CON'T)FIRE PREVENTION								
WELDING & CUTTING	1910.252							
(MULTIPLE REFERENCES)	1926.352							
FIRE PROTECTION (SEE ALSO FIRE PROTECTION EQUIPMENT)								
ACETYLENE							ISDS494	
AIRCRAFT INTERIOR		421						
AIRCRAFT INVESTIGATORS MANUAL		422M						
AIRCRAFT RAMP FIRE HAZARDS			Z215.1					
AIRCRAFT RESCUE - STANDARD OPERATING PROCEDURE		402						
AMMONIUM NITRATE	1910.109							
AUTOMATIC SPRINKLER SYSTEMS	1910.159							
BLASTING AGENTS	1910.109							
BORON HYDRIDES							ISDS508A	
BULK PLANTS	1910.106							
CARBON DIOXIDE SYSTEMS	1910.161	12	A54.1					
CHEMICAL PLANTS	1910.106							
CLEANING CONTAINERS FOR COMBUSTIBLES							ISDS432	
CLEANING PETROLEUM STORAGE TANK				RP-2015				
CODING SYSTEM FOR FIRE REPORTING		901						
COMBUSTIBLE METALS							ISDS567	
CUTOFFS	1926.150							
DEFINITION	1910.156							
DRY CHEMICAL SYSTEMS	1910.160	17						
EFFECTIVE DATES	1910.165							
EFFECTS OF FIRE ON OPERATIONS, MANAGEMENT		8						
EGRESS	1910.37							
ELECTRIC CUTTING EQUIPMENT - PETROLEUM OPERATION				RP-2009				
ELECTRIC SPARK & PETROLEUM PRODUCTS				RP-2003				
ELECTRIC WELDING EQUIPMENT - PETROLEUM OPERATION				RP-2009				
ELECTROSTATIC APPARATUS	1910.107							
ELECTROSTATIC DETEARING							ISDS468	
ELECTROSTATIC PAINT SPRAYING							ISDS468	
ENGINE FUELING (FIXED, PORTABLE, SELF-PROPELLED)				PSD-2203				
EXPLOSIVES	1910.109							
EXPOSURE FIRES		80A						
FIRE ATTACK TRAINING		197						
FIRE BRIGADES							ISDS588	
FIRE DEPARTMENT ORGANIZATION		4A						
FIRE EMERGENCY, MANAGEMENT		7						
FIRE FIGHTER - PROFESSIONAL QUALIFICATIONS		1001						
FIRE REPORTING FIELD INCIDENT MANUAL		901A						
FIRE-RESISTANT WATER-IN-OIL EMULS-HYDRAUL FLUID							ISDS543	
FIXED EQUIPMENT	1910.158							
(MULTIPLE REFERENCES)	1910.159							
(MULTIPLE REFERENCES)	1910.162							
FLAME-RETARDANT FABRICS							ISDS517A	
FLAMMABLE LIQUID	1910.106							
FLAMMABLE LIQUID IN SMALL CONTAINERS							ISDS532	
FLAMMABLE LIQUID TANK CAR CLEANING				APM-13				

SUBJECT	USDL & USDI	NFPA	ANSI	API	MCA		NSC	MISC
		APPLICABLE STANDARDS, GUIDES AND PUBLICATIONS						
(CON'T)FIRE PROTECTION (SEE ALSO FIRE PROTECTION EQUIPMENT)								
FLAMMABLE LIQUID TANK VEHICLE CLEANING				RP-2013				
FLOUR & FEED MILLS		61C						
FOAM EXTINGUISHING SYSTEMS		11						
FOAM-WATER SPRINKLER & SPRAY SYSTEMS		16						
FOREST FIRE CONTROL		295						
FRACTURING WELLS WITH FLAMMABLE LIQUID				PSD-2204				
GAS CUTTING EQUIPMENT IN PETROLEUM OPERATION				RP-2009				
GAS WELDING EQUIPMENT IN PETROLEUM OPERATION				RP-2009				
GASEOUS OXYGEN							ISDS472	
GUARD SERVICE - FIRE LOSS PREVENTION		601						
HALON 1211 SYSTEM		12B						
HALON 1301 SYSTEMS		12A						
HIGH EXPANSION FOAM SYSTEMS		11A						
HOSPITAL EMERGENCY PREPAREDNESS		3M						
HOT WORK REPAIRS TANK BOTTOM				PSD-2207				
IDENTIFICATION - FIRE HAZARDS OF MATERIALS		704M						
INDOOR STORAGE	1910.177							
INDUSTRIAL FIRE LOSS PREVENTION		6						
INDUSTRIAL PLANTS	1910.106							
LEGISLATION - MODEL DRAFTS		2M						
LIGHTNING & PETROLEUM PRODUCTS				RP-2003				
LOCAL FIRE ALARM SIGNALING SYSTEMS	1910.163							
LPG	1910.110							
MILLING AGRICULTURAL COMMODITIES		61D						
OIL REFINERY				RP-2001				
ORGANIZATION FOR FIRE SERVICES		4						
OXYGEN-ENRICHED ATMOSPHERE		53M						
PORTABLE EQUIPMENT	1910.158							
PORTABLE FIRE EXTINGUISHERS	1910.157							
PREVENTION PROGRAM	1926.24							
PROCESSING PLANTS	1910.106							
QUANTITY & LOCATION	75.1100-2							
RADAR HAZARDS							ISDS481	
REFINERIES	1910.106							
REQUIREMENTS	75.1100							
SAFETY					SG-17			
SALVAGING		604						
SERVICE STATIONS	1910.106			APM-5				
SPRAY BOOTHS	1910.107							
SPRAY FINISHING	1910.107							
STANDARDS SOURCES	1910.165							
STANDPIPE & HOSE SYSTEMS	1910.158							
STATIC ELECTRICITY & PETROLEUM PRODUCTS				RP-2003				
STORAGE AREAS	1910.177							
STORAGE TANK	1910.106							
SYNTHETIC FOAM & COMBINED AGENT SYSTEM		11B						
TRAINING REPORTS & RECORDS		9						
TRUCK	1910.178							
TRUCK - ELECTRIC BATTERY POWERED			B56.3					UL 583

APPLICABLE STANDARDS, GUIDES AND PUBLICATIONS

SUBJECT	USDL & USDI	NFPA	ANSI	API	MCA		NSC	MISC
(CON'T)FIRE PROTECTION (SEE ALSO FIRE PROTECTION EQUIPMENT)								
TRUCK - INTERNAL COMBUSTION ENGINE			B56.4					UL 558
TRUCK FIRE		512						
TYPE & QUALITY	75.1100-1							
WATER CHARGES - PRIVATE PROTECTION		292M						
WATER SUPPLY & EQUIPMENT	1926.150							
WETTING AGENTS		18						
FIRE PROTECTION EQUIPMENT (SEE ALSO FIRE PROTECTION)								
AIRCRAFT FOAM FIRE FIGHTING VEHICLE		412						
AIRCRAFT RESCUE		403						
(MULTIPLE REFERENCES)		414	B128.1					
(MULTIPLE REFERENCES)			Z213.1					
AUTOMATIC FIRE DETECTORS		72E						
AUTOMOTIVE FIRE APPARATUS		19						
CARBON DIOXIDE		12	A54.1					
COLOR CODING	1910.144							
EFFECTIVE DATES	1910.165							
EXTINGUISHER - CARBON DIOXIDE	1910.161							
EXTINGUISHER - CRANES & DERRICKS	1926.550							
EXTINGUISHER - DIP TANKS	1910.108							
EXTINGUISHER - FIXED DRY CHEMICAL	1910.160							
EXTINGUISHER - HAND, AIRCRAFT		408	Z112.3					
EXTINGUISHER - MOTOR VEHICLES	1926.600							
EXTINGUISHER - PORTABLE, CLASSIFICATION	1910.156							
EXTINGUISHER - PORTABLE, DISTRIBUTION	1910.157							
EXTINGUISHER - PORTABLE, GENERAL REQUIREMENTS	1910.157							
EXTINGUISHER - PORTABLE, HYDROSTATIC TEST INTERV	1910.157							
EXTINGUISHER - PORTABLE, INSPECT & MAINTENANCE	1910.157							
EXTINGUISHER - PORTABLE, INSTALLATION		10	Z112.1					
EXTINGUISHER - PORTABLE, MODEL ENABLING ACT		10L						
EXTINGUISHER - PORTABLE, SELECTION	1910.157							
EXTINGUISHER - RATING & FIRE TESTING			Z211.1					UL 711
EXTINGUISHER - SODA-ACID			Z171.1					UL 7
EXTINGUISHER - WELDING & CUTTING	1926.351							
FIRE ALARM SYSTEM - MUNICIPAL		73						
FIRE BRIGADES, PRIVATE		27						
FIRE DEPARTMENT - AIRCRAFT CRASH FIRES		406M						
FIRE DETECTION, EGRESS	1910.37							
FIRE DOORS & WINDOWS		80	A2.7					
FIRE FIGHTING EQUIPMENT	1926.150							
FIRE HOSE		196						
FIRE HOSE CARE		198						
FIRE HOSE COUPLINGS, SCREW THREAD		194						
FIRE HYDRANT - UNIFORM MARKINGS		291						
FIRE PUMP - CENTRIFUGAL		20						
FIRE PUMP - STEAM		21						
FIRE WARNING SYSTEM - HOUSEHOLD		74						
FIXED, AUTOMATIC SPRINKLERS	1910.159							
FIXED, CARBON DIOXIDE	1910.161							

SUBJECT	USDL & USDI	NFPA	ANSI	API	MCA		NSC	MISC
	APPLICABLE STANDARDS, GUIDES AND PUBLICATIONS							
(CON'T)FIRE PROTECTION EQUIPMENT (SEE ALSO FIRE PROTECTION)								
FIXED, DRY CHEMICALS	1910.160							
FLAME ARRESTOR, STORAGE TANK VENTS				PSD-2210				
FOAM EXTINGUISHERS, DIP TANKS	1910.108							
FOAM-WATER SPRINKLER & SPRAY SYSTEMS		16						
HYDROSTATIC TESTS	1910.157							
INSPECTION					SG-13			
LADDERS - FIRE DEPARTMENT		193						
MAINTENANCE	1910.160				SG-13			
MARKINGS	1910.157							
OUTSIDE PROTECTION		24						
PORTABLE, EXTINGUISHERS	1910.157							
PORTABLE, HOSE SYSTEMS	1910.158							
PORTABLE, STANDPIPE SYSTEMS	1910.158							
PUMPING UNITS, PORTABLE		191						
RECORDS	1910.157							
(MULTIPLE REFERENCES)	1910.160							
RECORDS - FIRE RESISTANCE CLASSIFICATION			Z134.1					UL 72
RESPIRATORY EQUIPMENT			Z88.5					
RESPIRATORY EQUIPMENT - FIREFIGHTERS		198						
SHIP FIRE SIGNAL		311						
SIGNAL BOX - MANUALLY ACTUATED			A154.1					UL 38
SMOKE DETECTOR - PHOTOELECTRIC			A131.4					UL 168
SPRINKLER SYSTEMS	1910.37							
(MULTIPLE REFERENCES)	1910.156							
(MULTIPLE REFERENCES)	1910.159							
SPRINKLER SYSTEMS - INSTALLATION		13	A54.2					
STANDARDS SOURCES	1910.165							
STANDPIPE & HOSE SYSTEMS		14						
TESTING	1910.157							
(MULTIPLE REFERENCES)	1910.160							
(MULTIPLE REFERENCES)	1910.161							
THERMOSTAT			Z220.1					UL 521
VALVE SUPERVISION		26						
VAPORIZING LIQUID AGENTS, HAZARDS		182M						
WATER SPRAY FIXED SYSTEMS		15	A54.2					
WATER SYSTEMS - RURAL FIRE PROTECTION		25						
FIRE RESISTANCE (RATING)								
STORAGE CABINETS	1910.106							
TANK SUPPORTS	1910.106							
FIRE RESISTANT								
BUILDING MATERIALS		703						
FABRIC TREATMENT							ISDS517A	
OIL-IN-WATER HYDRAULIC FLUID							ISDS543	
PAINTS	1910.37							
TEXTILES & FILMS - FIRE TEST		701						
TEXTILES - TEST METHOD			L14.69					ASTM D1230-61
(MULTIPLE REFERENCES)								AATCC 33

SUBJECT	USDL & USDI	NFPA	ANSI	API	MCA		NSC	MISC
	APPLICABLE STANDARDS, GUIDES AND PUBLICATIONS							
(CON'T)FIRE RESISTANT								
TRAILING CABLES	75.600							
FIRE SUPPRESSION EQUIP (SEE FIRE PROTECTION EQUIPMENT)								
FIRE TESTS								
BUILDING CONSTRUCTION & MATERIALS		251	A2.1					ASTM E119-69
DOOR ASSEMBLIES		252	A2.2					ASTM E152-66
FLAME-RESISTANT TEXTILES & FILMS		701						
ROOF COVERINGS		256						
SURFACE BURNING CHARACTERISTICS - BUILDING MTRLS		255						ASTM E84-68
WINDOW ASSEMBLIES		257	A2.6					ASTM E163-65
FIRE WATCH								
WELDING	1910.252							
FIREARMS								
PLANT PROTECTION							ISDS413A	
FIREPLACE								
FIRE CODE		211	A52.1					
SAFETY STANDARD			A131.3					UL 127
FIREWORKS								
MANUFACTURE		44A						
STORAGE		44A						
TRANSPORTATION		44A						
FIRST AID & HEALTH(SEE ALSO IND HYG &/OR MEDICAL)	1910.151							
(MULTIPLE REFERENCES)	1926.23							
(MULTIPLE REFERENCES)	1926.50							
(MULTIPLE REFERENCES)	1926.950							
API FIRST AID INSTRUCTOR TRAINING GUIDE				BULL-2019				
BLACK WIDOW SPIDER							ISDS258	
CHEMICAL BURNS							ISDS523	
FEDERAL SUPPLY CONTRACTS	50-204.5							
FIRST AID KIT	1915.58							
(MULTIPLE REFERENCES)	1916.58							
(MULTIPLE REFERENCES)	1917.58							
(MULTIPLE REFERENCES)	1918.96						ISDS202	
FIRST AID TRAINING (FLIP CHART)				PUB-2018				
FIRST AID TRAINING GUIDE				PUB-2017				
FLAMMABLE LIQUID TANK CAR CLEANING				APM-13				
INDUSTRIAL SKIN DISEASES							ISDS510	
INSPECTION	1915.58							
(MULTIPLE REFERENCES)	1916.58							
(MULTIPLE REFERENCES)	1917.58							
(MULTIPLE REFERENCES)	1918.96							
LABOR CAMPS	1910.142							
LOCATION	1915.58							
(MULTIPLE REFERENCES)	1916.58							

SUBJECT	USDL & USDI	NFPA	ANSI	API	MCA		NSC	MISC
	APPLICABLE STANDARDS, GUIDES AND PUBLICATIONS							
(CON'T)FIRST AID & HEALTH(SEE ALSO IND HYG &/OR MEDICAL)								
MARINE OPERATIONS	1926.605							
MATERIALS HANDLING					SG-1			
POISON IVY							ISDS304	
POISON OAK							ISDS304	
POISON SUMAC							ISDS304	
POLE TOP RESUSCITATION							ISDS376	
RESPIRATORY PROTECTION - HOME THERAPY		56HM						
RESPIRATORY PROTECTION EQUIPMENT							ISDS444A	
TICK BITES							ISDS228	
WORK ACCIDENT RECORDS & ANALYSIS							ISDS527A	
FISSIONABLE MATERIAL	1915.37							
(MULTIPLE REFERENCES)	1916.37							
HANDLING OUTSIDE NUCLEAR REACTOR			N16.1					
SYMBOL			N12.1					
FITTING-UP	1926.752							
FIXED BARRIER GUARDS	1910.211							
FIXED DRY CHEMICAL EXTINGUISHER SYSTEMS								
ALARMS & INDICATORS	1910.160							
GENERAL REQUIREMENTS	1910.160							
INSPECTION & MAINTENANCE	1910.160							
FIXED INDUSTRIAL STAIRS	1910.24							
ANGLE OF STAIRWAYS	1910.24							
APPLICATION OF REQUIREMENTS	1910.24							
GENERAL	1910.24							
LENGTH	1910.24							
OPEN RISERS	1910.24							
RAILINGS & HANDRAILS	1910.24							
STRENGTH	1910.24							
THREADS	1910.24							
VERTICAL CLEARANCE	1910.24							
WHERE REQUIRED	1910.24							
WIDTH	1910.24							
FIXED LADDERS	1910.27							
CAGES OR WELLS	1910.27							
CLEARANCE	1910.27							
CLEARANCE CLIMBING SIDE	1910.27							
CLEARANCE HATCH COVERS	1910.27							
CLEARANCE IN BACK	1910.27							
CLEARANCE OF GRAB BAR	1910.27							
CLEARANCE STEP-ACROSS DISTANCE	1910.27							
CLEARANCE WITH CAGES	1910.27							
CLEARANCE WITHOUT CAGES	1910.27							
DESIGN REQUIREMENTS	1910.27							

APPLICABLE STANDARDS, GUIDES AND PUBLICATIONS

SUBJECT	USDL & USDI	NFPA	ANSI	API	MCA		NSC	MISC
(CON'T)FIXED LADDERS								
EXTENSIONS	1910.27							
LANDING PLATFORMS	1910.27							
MAINTENANCE	1910.27							
PITCH	1910.27							
REQUIREMENTS	1910.27							
SPECIFIC FEATURES	1910.27							
FIXED MACHINERY ANCHORS	1910.212							
FLAME ARRESTOR								
STORAGE TANK VENT - PETROLEUM OIL/GASOLINE			Z222.1					UL 525
FLAMMABILITY								
TEXTILES - TEST METHOD			L14.69					ASTM D1230-61
(MULTIPLE REFERENCES)								AATCC 33
WEARING APPAREL		702						
FLAMMABLE & COMBUSTIBLE GASES								
PROPERTIES		325M						
FLAMMABLE & COMBUSTIBLE LIQUIDS	1910.106							
(MULTIPLE REFERENCES)	1926.152							
BULK PLANTS	1910.106							
CLASSIFICATION	1910.106	321						
CLEANING SMALL TANKS		327						
COATINGS, ORGANIC - MANUFACTURE		35						
CONTAINER & PORTABLE TANK STORAGE	1910.106							
DEFINITION	1910.106							
DEMOLITION	1926.850							
DIP TANKS	1910.108	34						
DISPENSER			Z216.1					UL 87
DRY CLEANING PLANT		32	Z8.2					
FARM STORAGE		395						
FIRE CODE		30						
FIRE PROTECTION	1910.106							
GASOLINE BLOW TORCH		393						
HOT WORK	1915.34							
(MULTIPLE REFERENCES)	1916.34							
(MULTIPLE REFERENCES)	1917.34							
HOUSEKEEPING	1910.106							
IN SMALL CONTAINER							ISDS532	
INDUSTRIAL PLANTS	1910.106							
LEAKAGE - UNDERGROUND STORAGE TANK		329						
LIQUID TRANSFER	1910.107							
MANHOLES		328						
MATERIALS HANDLING					SG-3			
OPEN FLAMES IN MANHOLES	1926.956							
PAINTING	1915.24							
(MULTIPLE REFERENCES)	1916.24							

SUBJECT	USDL & USDI	NFPA	ANSI	API	MCA		NSC	MISC
	APPLICABLE STANDARDS, GUIDES AND PUBLICATIONS							
(CON'T)FLAMMABLE & COMBUSTIBLE LIQUIDS								
PIPE CONNECTORS			B148.1					UL 567
PIPELINE WELDING/HOT TAPPING				PSD-2201				
PIPING, VALVES & FITTINGS	1910.106							
PRECAUTIONS	1915.25							
(MULTIPLE REFERENCES)	1916.25							
PROCESSING PLANTS	1910.106							
PROPERTIES		325M						
REFINERIES, CHEMICAL PLANTS & DISTILLERIES	1910.106							
SAFEGUARDING EMPLOYEES	1915.21							
(MULTIPLE REFERENCES)	1916.21							
SAFETY RELIEF DEVICES	1910.107							
SERVICE STATIONS	1910.106							
SEWERS		328						
SHIPPING TANKS, PORTABLE		386						
SMOKING	1910.106							
SOLVENT EXTRACTION		36						
SOLVENTS	1915.21							
(MULTIPLE REFERENCES)	1916.21							
SPRAY FINISHING	1910.107	33						
STORAGE	1910.106				SG-3			
STORAGE & HANDLING - DIP TANKS	1910.108							
TANK CAR CLEANING				APM-13				
TANK CAR UNLOADING					TC-4			
TANK VEHICLE		385						
TANK VEHICLE CLEANING				RP-2013				
TANK WELDING/HOT TAPPING				PSD-2201				
VENTILATION	1910.108							
VESSEL WELDING/HOT TAPPING				PSD-2201				
WELDING	1915.31							
(MULTIPLE REFERENCES)	1916.31							
FLAMMABLE & COMBUSTIBLE SOLIDS								
PROPERTIES		325M						
FLAMMABLE & TOXIC SUBSTANCES								
COLD WORK	1915.12							
(MULTIPLE REFERENCES)	1917.12							
HOT WORK	1915.13							
(MULTIPLE REFERENCES)	1917.13							
HOUSEKEEPING	1915.51							
(MULTIPLE REFERENCES)	1916.51							
(MULTIPLE REFERENCES)	1917.51							
FLANGES, ABRASIVE WHEELS	1910.215							

SUBJECT	USDL & USDI	NFPA	ANSI	API	MCA		NSC	MISC
	APPLICABLE STANDARDS, GUIDES AND PUBLICATIONS							
FLASHLIGHT								
VAPOR IGNITION RISK				PSD-2212				
FLEXIBLE INSULATED PROTECTIVE EQUIPMENT								
ELECTRICAL WORKER							ISDS598	
FLOATS								
BULL	1926.700							
SCAFFOLDS	1926.451							
FLOODING, TANK AREAS	1910.106							
FLOORS, FLOORING	1910.23						ISDS595	
COMBUSTIBILITY OF SWEEPING COMPOUNDS			Z240.1					UL 41
DRAINING		92M						
ELECTRICALLY CONDUCTIVE			A148.1					UL 779
FALLS ON FLOOR							ISDS495	
FLOOR OPENINGS, SKYLIGHT	1910.23							
FLOOR OPENINGS, TEMPORARY	1910.23							
GENERAL REQUIREMENTS	1910.22							
HINGED COVERS	1910.23							
MANLIFTS	1910.68							
MATS & RUNNERS							ISDS595	
OPENINGS	1926.500		A12.1					
OPENINGS, COVERS	1910.23							
OPENINGS, PROTECTION	1910.23							
SLIPPERY							ISDS286	
STEEL ERECTION	1926.750							
TRAPDOORS	1910.23							
WATERPROOFING		92M						
FLOOR LOADING PROTECTION	1910.22							
FLOUR & FEED MILLS								
DUST EXPLOSION		61C	Z12.3					
FIRE PREVENTION		61C						
FLOW COATINGS	1910.108							
FLUORESCENCE ANALYSIS								
RADIATION SAFETY			N43.2					
FLUORIDES								
INORGANIC							ISDS442	
FLUORIDE BEARING DUSTS & FUMES								HG

SUBJECT	USDL & USDI	NFPA	ANSI	API	MCA		NSC	MISC
	APPLICABLE STANDARDS, GUIDES AND PUBLICATIONS							
FLUORIDE DUST - INORGANIC								
TOXIC EFFECT			Z37.28					
FLUORINE								HG
FLUORINE (ATMOSPHERIC)								
AIR QUALITY CRITERIA & VEGETATION				AQM 69-3				
FLUORINE (LIQUID)					CC-10			
(MULTIPLE REFERENCES)					TEG			
FLUOROCARBON 11								
TOXIC EFFECT			Z37.36					
FLUOROCARBON 12								
TOXIC EFFECT			Z37.38					
FLUOROCARBON 113								
TOXIC EFFECT			Z37.35					
FLUOROSIS								
LIVESTOCK				AQM 69-4				
FLYWHEEL GUARDING	1910.219							
FOOD HANDLING	1926.51							
LABOR CAMPS	1910.141							
FOOT PROTECTION	1910.136							
(MULTIPLE REFERENCES)	1915.83							
(MULTIPLE REFERENCES)	1916.83							
(MULTIPLE REFERENCES)	1917.83							
(MULTIPLE REFERENCES)	1918.104		Z41.1					
FOOTING EXCAVATIONS	1926.955							
(MULTIPLE REFERENCES)	1926.957							
FOOTWALKS, OVERHEAD GANTRY CRANES	1910.179							
FORGING								
AIR LIFT HAMMER	1910.218							
BENDING, FORGING MACHINES	1910.211							
FORGING HAMMER DIES							ISDS467	
FORGING PRESS, MECHANICAL							ISDS560	
RECORDS	1910.218							
SAFETY STANDARD			B24.1					

SUBJECT	APPLICABLE STANDARDS, GUIDES AND PUBLICATIONS							
	USDL & USDI	NFPA	ANSI	API	MCA		NSC	MISC
FORKLIFT					SG-6			
FORM, FORMS, FORMWORK								
CONCRETE	1926.700							
CONCRETE - TIMBER, TIMBERING	1926.701							
SCAFFOLDS	1926.451							
FORMS, DOCUMENTATION	1919.90							
FORMALDEHYDE					CIC-48			
(MULTIPLE REFERENCES)					SD-1		ISDS342	HG
TOXIC EFFECT			Z37.16					
PARA-FORMALDEHYDE					SD-6			
FORMALIN (SEE FORMALDEHYDE)								
FORMIC ACID					CC-83			
(MULTIPLE REFERENCES)					CIC-49			
(MULTIPLE REFERENCES)					TEG			
FRACTURING								
OIL/GAS WELL WITH FLAMMABLE LIQUID				PSD-2204				
FRAMES								
ELECTRICALLY ILLUMINATED			C33.19					UL 368
FREEZER								
HOME - SAFETY STANDARD			B97.1					UL 250
FREIGHT CAR								
DOORS							ISDS374	
FREQUENCY								
ACOUSTICAL MEASUREMENTS			S1.6					
FRONT-END LOADER							ISDS589	
FUEL, FUELING & REFUELING								
AIRCRAFT		407	Z119.1					
CRANES & DERRICKS	1926.550							
ENGINES (FIXED, PORTABLE, SELF-PROPELLED)				PSD-2203				
FLAMMABLE & COMBUSTIBLE LIQUIDS	1926.152							
FUEL GASES	1926.350							
FUEL POWERED TOOLS	1926.302							
LPG DISPENSING	1926.153							
OXYGEN MANIFOLDS	1926.350							

SUBJECT	USDL & USDI	NFPA	ANSI	API	MCA		NSC	MISC
	APPLICABLE STANDARDS, GUIDES AND PUBLICATIONS							
FUEL GAS SYSTEMS	1910.252							
PIPING			B31.2					
FUEL STORAGE, INDUSTRIAL TRUCKS	1910.178							
FULMINATE OF MERCURY							ISDS309A	
FUMES (SEE ALSO EXHAUST SYSTEMS &/OR VENTILATION)								
AIR CONTAMINANTS	1910.93							
(MULTIPLE REFERENCES)	1910.178							
FEDERAL SUPPLY CONTRACTS	50-204.50							
IN INDUSTRY							ISDS531	
VENTILATION	1926.55							
FUMIGATION								
CARGO	1918.93							
GAS SYSTEM		57						
GRAIN	1918.93							
STOWAGE SPACES	1918.93							
TOBACCO	1918.93							
FURFURAL					CIC-50			HG
FURNACE								
COAL-FIRED		85E						
FUEL OIL-FIRED		85D						
GAS-FIRED WATERTUBE		85						
INDUSTRIAL		86B						
NATURAL GAS-FIRED - EXPLOSION		85B						
OIL-FIRED WATERTUBE		85						
OVENS & FURNACES		86A						
SPECIAL PROCESSING ATMOSPHERE		86C						
FUSE								
SAFETY	1926.907							
STORAGE & HANDLING							ISDS358	
FUSEHOLDER								
SAFETY STANDARD			C33.10					UL 512
FUSIBLE PLUGS								
CYLINDERS	1910.167							
TANKS	1910.168							

O C C U P A T I O N A L   S A F E T Y ,   H E A L T H   &   F I R E   I N D E X

SUBJECT	USDL & USDI	NFPA	ANSI	API	MCA		NSC	MISC
	APPLICABLE STANDARDS, GUIDES AND PUBLICATIONS							
GAGING DEVICES	1910.110							
GAMMA-RAY								
DOSIMETER			N13.5					
RADIATION SURVEY INSTRUMENT			N13.4					
SAFETY STANDARD - NON-MEDICAL SOURCE			Z54.1					NBS H93
GANGWAYS	1910.21							
ACCESS, FLOATING DRYDOCKS	1915.45							
(MULTIPLE REFERENCES)	1916.45							
(MULTIPLE REFERENCES)	1917.45							
ACCESS, VESSELS	1915.44							
(MULTIPLE REFERENCES)	1916.44							
(MULTIPLE REFERENCES)	1917.44							
COAST GUARD INSPECTED	1918.11							
DEFINITION	1915.2							
(MULTIPLE REFERENCES)	1916.2							
(MULTIPLE REFERENCES)	1917.2							
(MULTIPLE REFERENCES)	1918.3							
DUCKBOARDS	1918.21							
GENERAL REQUIREMENTS	1918.11							
ILLUMINATION	1915.44							
(MULTIPLE REFERENCES)	1916.44							
(MULTIPLE REFERENCES)	1917.44							
(MULTIPLE REFERENCES)	1918.21							
INDUSTRIAL SAFETY DATA SHEET							ISDS318	
LOADS	1918.21							
NETS	1918.21							
OBSTRUCTIONS	1915.44							
(MULTIPLE REFERENCES)	1916.44							
(MULTIPLE REFERENCES)	1917.44							
(MULTIPLE REFERENCES)	1918.21							
RAIL HEIGHTS	1915.45							
(MULTIPLE REFERENCES)	1916.45							
(MULTIPLE REFERENCES)	1917.45							
RAILINGS	1918.11							
STANCHIONS	1918.11							
SUBSTITUTES	1918.11							
SUPPORTING BRIDLES	1915.44							
(MULTIPLE REFERENCES)	1916.44							
(MULTIPLE REFERENCES)	1917.44							
(MULTIPLE REFERENCES)	1918.21							
TRIMMING	1918.21							
TURNTABLES	1918.11							
WALKING SURFACES	1918.11							
WALKWAY WIDTH	1915.45							
(MULTIPLE REFERENCES)	1916.45							
(MULTIPLE REFERENCES)	1917.45							

APPLICABLE STANDARDS, GUIDES AND PUBLICATIONS

SUBJECT	USDL & USDI	NFPA	ANSI	API	MCA		NSC	MISC
GANTRY TRUCKS							ISDS257	
GARAGE								
MECHANIZED - SAFETY CODE			A113.1					
PARKING		88A						
REPAIRS		88B						
UNDERCOATING OPERATIONS	1910.107							
GARBAGE	1918.94							
GARNETT MACHINES	1910.262							
INDUSTRIAL SAFETY DATA SHEET							ISDS366A	
GAS	1926.55							
FEDERAL SUPPLY CONTRACTS	50-204.50							
FUEL GASES	1926.350							
GASEOUS CONDITIONS	1926.651							
GAS, COMPRESSED	1910.101							
(MULTIPLE REFERENCES)	1910.166							
COMPRESSED GAS CYLINDERS	1910.166							
CYLINDER INSPECTION	1910.101							
(MULTIPLE REFERENCES)	1910.166							
FUMES CONTROL, GENERAL	1910.93							
FUMES CONTROL, INDUSTRIAL TRUCKS	1910.178							
LPG	1910.110							
MASKS	1910.134							
VAPORS, FUMES, DUSTS, MISTS	1910.93							
GAS, FLAMMABLE & COMBUSTIBLE								
PROPERTIES		325M						
GAS CUTTING EQUIPMENT								
SAFETY & FIRE PROTECTION IN PETROLEUM OPERATIONS				RP-2009				
GAS DRILLING								
SAFETY				RP-2011				
GAS EQUIPMENT								
BOILER, USED IN			Z83.3					
INSTALLATION		54	Z21.30					
(MULTIPLE REFERENCES)			Z83.1					
SAFETY CONTROLS			Z226.1					UL 372
GAS FREE CONDITIONS								
HAZARDOUS RESIDUES	1915.14							
(MULTIPLE REFERENCES)	1917.14							
HOT WORK	1915.14							
(MULTIPLE REFERENCES)	1917.14							
INFORMING CREW	1915.14							
(MULTIPLE REFERENCES)	1917.14							

SUBJECT	USDL & USDI	NFPA	ANSI	API	MCA		NSC	MISC
	APPLICABLE STANDARDS, GUIDES AND PUBLICATIONS							
(CON'T)GAS FREE CONDITIONS								
SPACES	1915.14							
(MULTIPLE REFERENCES)	1917.14							
GAS HEATER								
FORCED AIR HEATER			Z83.5					
GAS MASK								
CANISTER IDENTIFICATION			K13.1					
COLOR CODES	1910.134							
LABELING	1910.134							
GAS SHIELDED ARC							ISDS552	
GAS SYSTEM								
ACETYLENE CYLINDER CHARGING PLANT		51A						
BULK OXYGEN SYSTEMS		50						
COMBUSTION ENGINE, STATIONARY		37						
CUTTING PROCESSES		51B						
FUMIGATION		57						
GAS TURBINE		37						
GASEOUS HYDROGEN SYSTEMS		50A						
HYPERBARIC FACILITIES		56D						
HYPOBARIC FACILITIES		56E						
INHALATION ANESTHETICS		56A						
INHALATION THERAPY		56B						
LIQUEFIED HYDROGEN SYSTEMS		50B						
LNG STORAGE & HANDLING		59A						
LPG AT UTILITY GAS PLANTS		59						
LPG STORAGE & HANDLING		58						
NONFLAMMABLE MEDICAL SYSTEMS		56F						
OXYGEN-FUEL GAS SYSTEM FOR CUTTING		51						
OXYGEN-FUEL GAS SYSTEM FOR WELDING		51						
WELDING PROCESSES		51B						
GAS TRANSMISSION								
PIPING			B31.8					
PIPING - INSTALLATION		54	Z21.30					
(MULTIPLE REFERENCES)			Z83.1					
GAS TURBINE								
GAS SYSTEM		37						
GAS WELDING & CUTTING								
FUEL GASES	1915.35							
(MULTIPLE REFERENCES)	1916.35							
(MULTIPLE REFERENCES)	1917.35							
HOSE	1915.35							
(MULTIPLE REFERENCES)	1916.35							
(MULTIPLE REFERENCES)	1917.35							

SUBJECT	USDL & USDI	NFPA	ANSI	API	MCA		NSC	MISC
	APPLICABLE STANDARDS, GUIDES AND PUBLICATIONS							
(CON'T)GAS WELDING & CUTTING								
MANIFOLDS	1915.35							
(MULTIPLE REFERENCES)	1916.35							
(MULTIPLE REFERENCES)	1917.35							
PLACEMENT & TREATMENT OF CYLINDERS	1915.35							
(MULTIPLE REFERENCES)	1916.35							
(MULTIPLE REFERENCES)	1917.35							
PRESSURE REGULATORS	1915.35							
(MULTIPLE REFERENCES)	1916.35							
(MULTIPLE REFERENCES)	1917.35							
SAFETY & FIRE PROTECTION IN PETROLEUM OPERATIONS				RP-2009				
TORCHES	1915.35							
(MULTIPLE REFERENCES)	1916.35							
(MULTIPLE REFERENCES)	1917.35							
TRANSPORTING, MOVING & STORING CYLINDERS	1915.35							
(MULTIPLE REFERENCES)	1916.35							
(MULTIPLE REFERENCES)	1917.35							
GAS WELL								
FRACTURING WITH FLAMMABLE LIQUID				PSD-2204				
GASEOUS HYDROGEN SYSTEMS (SEE HYDROGEN SYSTEMS)								
GASEOUS OXYGEN							ISDS472	
GASOLINE				TR				
FLAME ARRESTOR - STORAGE TANK VENT			Z222.1					UL 525
GASOLINE, LEADED								
STORAGE TANK TOXIC EFFECT				PSD-2202				
GASOLINE ENGINE								
COLD-ROOM TESTING							ISDS465	
GASOLINE STATIONS	1910.106							
GAUGING								
PETROLEUM TANKS							ISDS563	
GEAR								
EXAMINATIONS	1919.31							
HEAT TREATMENT	1919.80							
INSPECTION	1915.61							
(MULTIPLE REFERENCES)	1916.61							
(MULTIPLE REFERENCES)	1917.61							
PROOF TESTS	1919.31							
SAFE WORKING LOADS	1915.61							
(MULTIPLE REFERENCES)	1916.61							
USE	1915.66							
(MULTIPLE REFERENCES)	1916.66							
(MULTIPLE REFERENCES)	1917.66							

SUBJECT	USDL & USDI	NFPA	ANSI	API	MCA		NSC	MISC
	APPLICABLE STANDARDS, GUIDES AND PUBLICATIONS							
GEAR, OTHER THAN SHIP								
BLOCKS, SNATCH	1918.84							
BRIDLES	1918.67							
(MULTIPLE REFERENCES)	1918.81							
CHAINS	1918.64							
CLAMPS, BEAM OR FRAME	1918.84							
HOOKS	1918.66							
HOOKS, CASE	1918.66							
(MULTIPLE REFERENCES)	1918.81							
HOOKS, PLATE	1918.66							
INSPECTION	1918.61							
MANILA ROPE	1918.62							
MARKING WEIGHT	1918.61							
PALLETS	1918.67							
SHACKLES	1918.65							
STEVEDORE GEAR	1918.61							
SYNTHETIC FIBER ROPES	1918.62							
WIRE ROPE	1918.63							
WIRE ROPE CLIPS	1918.63							
WIRE ROPE LIMITATIONS	1918.63							
GEAR, SHIP CARGO HANDLING								
CARGO HOOKS	1918.52							
CERTIFICATION	1918.12							
COAMING ROLLERS	1918.52							
FALLS	1918.52							
HEEL BLOCKS	1918.52							
PREVENTERS	1918.52							
(MULTIPLE REFERENCES)	1918.54							
SAFE WORKING LOADS	1918.51							
STOPPERS, CHAIN	1918.52							
STOPPERS, CLAMP	1918.52							
UNSAFE	1918.51							
WIRE ROPE LIMITATIONS	1918.51							
GEARS, POWER TRANSMISSION	1910.219							
GEAR CERTIFICATION								
BRAKING DEVICES	1919.22							
CHAINS	1919.25							
COAST GUARD INSPECTED VESSELS	1918.12							
COMPETENT PERSONS	1919.37							
CONTAINER CRANES	1918.14							
DAMAGED COMPONENTS	1919.20							
DERRICK ATTACHMENT	1919.23							
FOREIGN VESSELS	1918.12							
GRACE PERIODS	1919.18							
HEAT TREATMENT	1919.16							
(MULTIPLE REFERENCES)	1919.17							
IRON ANNEALING	1919.36							

APPLICABLE STANDARDS, GUIDES AND PUBLICATIONS

SUBJECT	USDL & USDI	NFPA	ANSI	API	MCA		NSC	MISC
(CON'T)GEAR CERTIFICATION								
JURISDICTION TRANSFER	1918.13							
MAINTENANCE	1918.13							
MATERIALS HANDLING DEVICES	1918.13							
POWER SOURCES	1919.22							
PUBLIC VESSELS	1918.12							
REGULATIONS	1918.12							
ROPE, WIRE	1919.24							
ROUTINE DISASSEMBLY	1918.13							
SAFE WORKING LOADS	1919.13							
(MULTIPLE REFERENCES)	1919.21							
SEASONAL OPERATIONS	1918.13							
STANDARDS	1918.13							
TESTING	1919.14							
(MULTIPLE REFERENCES)	1919.15							
WELDING	1919.19							
GEAR-HOBBING MACHINE							ISDS362	
GENERAL ENVIRONMENTAL CONTROLS								
NONWATER CARRIAGE DISPOSAL SYSTEMS	1910.143							
SAFETY COLOR CODES FOR PHYSICAL HAZARDS	1910.144							
SANITATION	1910.141							
SPECIFICATIONS - ACCIDENT PREVENT SIGNS & TAGS	1910.145							
TEMPORARY LABOR CAMPS	1910.142							
GENERAL STORAGE								
BUILDING SERVICE EQUIPMENT	1910.177							
DEFINITION	1910.177							
FIRE PROTECTION REQUIREMENTS	1910.177							
GENERAL REQUIREMENTS	1910.177							
MECHANICAL HANDLING EQUIPMENT	1910.177							
PILING PROTECTION REQUIREMENTS	1910.177							
SMOKING	1910.177							
GENERATORS								
ACETYLENE	1910.252							
FOR HAZARDOUS USE			C33.72					
(MULTIPLE REFERENCES)			C33.93					UL 674A
(MULTIPLE REFERENCES)								UL 674B
GAS ATMOSPHERE			Z83.2					
GEOPHYSICAL EXPLORATION				BULL-2024				
GLAZING MATERIAL								
BUILDINGS			Z97.1					
MOTOR VEHICLE			Z26.1					

SUBJECT	USDL & USDI	NFPA	ANSI	API	MCA		NSC	MISC
	APPLICABLE STANDARDS, GUIDES AND PUBLICATIONS							
GLOVES	1926.951							
RUBBER INSULATING	1910.137							
GLYCERIN (SEE GLYCEROL)								
GOGGLES	1926.102							
LASERS	1926.102							
WELDING	1926.102							
GRADERS							ISDS256A	
GRADES, ROADWAY	1926.602							
GRAIN ELEVATOR								
DUST EXPLOSION		61B	Z12.4					
GRAIN FITTING								
DEEP TANKS, PROXIMITY	1918.98							
FEEDERS	1918.98							
HATCH COVER REMOVAL	1918.98							
INTERMEDIATE DECKS	1918.98							
LONGITUDINAL BULKHEADS OR SHIFTING BOARDS	1918.98							
REGULATIONS	1918.98							
WORK SURFACES, TEMPORARY	1918.98							
GRANDSTAND								
SAFETY STANDARD		102	Z20.3					
GRAPHIC ARTS								
PRESS			B65.1					
GRAVITY HAMMERS	1910.218							
GRAVITY ROLLERS	1918.68							
GRINDING, POLISHING & BUFFING								
BRANCH PIPES	1910.94							
DEFINITION	1910.94							
ENCLOSURE DESIGN	1910.94							
EXHAUST SYSTEM	1910.94							
HOODS	1910.94							
GRINDING MACHINES	1910.215							
(MULTIPLE REFERENCES)	1926.303							
CUTOFF WHEELS	1910.243							
FOOD	1910.263							
FORGING EQUIPMENT	1910.218							
GUARDING	1910.215							
PORTABLE							ISDS583	
SURFACE	1910.215							

SUBJECT	USDL & USDI	NFPA	ANSI	API	MCA		NSC	MISC
	APPLICABLE STANDARDS, GUIDES AND PUBLICATIONS							
(CON'T)GRINDING MACHINES								
TOP	1910.215							
VENTILATION	1910.94							
VENTILATION CONTROL			Z43.1					
GROUND LINES	1926.551							
GROUND SUPPORT								
EXCAVATIONS	1926.651							
TUNNELS	1926.800							
GROUNDING & BONDING	1918.76							
(MULTIPLE REFERENCES)	1926.954							
AIR MOVING DEVICES	1915.12							
(MULTIPLE REFERENCES)	1917.12							
ARC WELDING	1915.36							
(MULTIPLE REFERENCES)	1916.36							
(MULTIPLE REFERENCES)	1917.36							
ATTACHING GROUNDS	1926.954							
BULK OXYGEN SYSTEMS	1910.104							
CIRCUITS	1910.314							
COMBUSTIBLE LIQUIDS	1910.107							
ELECTRIC CRANES							ISDS287	
ELECTRIC EQUIPMENT - PORTABLE	1915.72							
(MULTIPLE REFERENCES)	1916.72							
(MULTIPLE REFERENCES)	1917.72							
ELECTRIC MOBILE EQUIPMENT							ISDS287	
ELECTRIC SHOVELS							ISDS287	
ELECTRICAL EQUIPMENT - PORTABLE							ISDS299	
ELECTRICAL SPRAY BOOTHS	1910.107							
ELECTROSTATIC SPRAYING	1910.107							
GROUND LEADS	1926.954							
GROUNDING & BONDING	1926.401							
GROUNDING ELECTRODES	1926.954							
HAND SPRAYING	1910.107							
IGNITION SOURCES	1910.107							
JUMPER CABLES	1926.550							
LIQUEFIED HYDROGEN SYSTEMS	1910.103							
LIQUID TRANSFER	1910.107							
METHODS	1910.314							
(MULTIPLE REFERENCES)	75.700							
NEW CONSTRUCTION	1926.954							
PAINT REMOVERS	1915.23							
REMOVING GROUNDS	1926.954							
RETURN & MACHINE GROUNDING	1926.351							
SAFETY STANDARD			C33.8					UL 467
SOLVENT DRUMS	1915.24							
(MULTIPLE REFERENCES)	1916.24							
SPRAYING OPERATIONS	1910.107							
(MULTIPLE REFERENCES)	1915.24							
(MULTIPLE REFERENCES)	1916.24							

APPLICABLE STANDARDS, GUIDES AND PUBLICATIONS

SUBJECT	USDL & USDI	NFPA	ANSI	API	MCA		NSC	MISC
(CON'T) GROUNDING & BONDING								
TEMPORARY LIGHTS	1915.52							
(MULTIPLE REFERENCES)	1916.52							
(MULTIPLE REFERENCES)	1917.52							
TESTING, VOLTAGE	1926.954							
TESTING, WITHOUT GROUNDS	1926.954							
TOWER	1926.954							
WELDING EQUIPMENT	1910.252							
WIRE CONDUCTORS	1926.954							
WOODWORKING TOOLS	1910.243							
GUARDING								
ABRASIVE WHEELS, MACHINERY	1910.215							
(MULTIPLE REFERENCES)	1915.74							
(MULTIPLE REFERENCES)	1916.74							
(MULTIPLE REFERENCES)	1917.74							
ABRASIVE WHEELS, PORTABLE	1910.243							
APPROVED MATERIALS	1910.219							
BARRELS	1910.212							
BELTS, ROPE & CHAIN	1910.219							
BENCH & FLOOR STANDS	1910.215							
CHAINS	1910.219							
CLUTCHES, CUTOFFS, COUPLINGS, PULLEYS	1910.219							
DEFINITION	1910.211							
DISKS, SHIELDS & "U" GUARDS	1910.219							
EXPLOSIVE ACTUATED TOOLS	1910.243							
FEDERAL SUPPLY CONTRACTS	50-204.5							
FLOOR OPENINGS	1910.23							
FLYWHEEL	1910.219							
FORGING MACHINES	1910.218							
FRICTION DRIVES	1910.219							
FRICTION DRIVES, MECHANICAL POWER TRANSMISSION	1910.219							
FRICTION DRIVES, SHAFTS	1910.219							
FRICTION DRIVES, STANDARD GUARDS	1910.219							
GEARS	1910.219							
GENERAL REQUIREMENTS FOR ALL MACHINES	1910.212							
GRINDERS	1910.215							
HORIZONTAL OVERHEAD BELTS	1910.219							
INDUSTRIAL TRUCKS	1910.178							
INFRARED HEATING LAMPS	1915.53							
(MULTIPLE REFERENCES)	1916.53							
KEYS, SET SCREWS	1910.219							
LADDERWAY	1910.23							
LAUNDRY MACHINERY	1910.264							
MACHINERY	1910.212							
MANHOLES	1926.956							
MATERIALS HANDLING AREAS	1910.176							
MECHANICAL POWER PRESSES	1910.217							
MILLS & CALENDERS IN THE RUBBER & PLASTIC IND	1910.216							
MOVING MACHINERY ON DRYDOCKS	1915.71							
(MULTIPLE REFERENCES)	1916.71							
(MULTIPLE REFERENCES)	1917.71							

SUBJECT	USDL & USDI	NFPA	ANSI	API	MCA		NSC	MISC
			APPLICABLE STANDARDS, GUIDES AND PUBLICATIONS					
(CON'T) GUARDING								
PNEUMATIC POWER TOOLS	1910.243							
PORTABLE ABRASIVE WHEELS	1910.243							
PORTABLE CIRCULAR SAWS	1915.71							
(MULTIPLE REFERENCES)	1916.71							
(MULTIPLE REFERENCES)	1917.71							
POWDER ACTUATED TOOLS	1915.75							
(MULTIPLE REFERENCES)	1916.75		A10.3					
POWER LAWN MOWERS	1910.243		B71.1					
POWER PRESS	1910.217							
POWER TOOLS, HAND & PORTABLE	1910.243							
PRIME-MOVER (FLYWHEELS)	1910.219							
STAIRWAYS & RAILINGS	1910.23							
TEMPORARY LIGHTS	1915.52							
(MULTIPLE REFERENCES)	1916.52							
(MULTIPLE REFERENCES)	1917.52							
WALL OPENINGS	1910.23							
WOOD	1910.219							
WOODWORKING MACHINERY REQUIREMENTS	1910.213							
WOODWORKING TOOLS, PORTABLE	1910.243							
GUARDING FLOOR & WALL OPENINGS & HOLES								
MATERIALS HANDLING & STORAGE	1910.176							
PROTECTION FOR FLOOR OPENINGS	1910.23							
PROTECTION FOR WALL OPENINGS & HOLES	1910.23							
PROTECTION OF OPEN-SIDED FLOOR, PLATFORM & RUNWAY	1910.23							
RAILINGS, TOEBOARDS & COVER SPECIFICATIONS	1910.23							
STAIRWAY RAILINGS & GUARDS	1910.23							
GUARDRAILS	1926.500							
CONCRETE MIXERS	1926.700							
COVERS & GUARDRAILS	1910.22							
CRANES & DERRICKS	1926.550							
DECK OPENINGS	1915.43							
(MULTIPLE REFERENCES)	1916.43							
ENGINE ROOM	1910.219							
GANGWAYS	1915.44							
(MULTIPLE REFERENCES)	1916.44							
(MULTIPLE REFERENCES)	1917.44							
MANHOLES	1915.43							
(MULTIPLE REFERENCES)	1916.43							
OPENINGS	1915.43							
(MULTIPLE REFERENCES)	1916.43							
RAILINGS	1910.24							
SAFETY STANDARD			A12.1					
SCAFFOLDING	1926.451							
SCAFFOLDS, STAGING, RUNWAYS, PLATFORMS	1915.41							
(MULTIPLE REFERENCES)	1916.41							
STAIRWAYS	1910.21							
(MULTIPLE REFERENCES)	1910.24							
(MULTIPLE REFERENCES)	1926.500							
(MULTIPLE REFERENCES)	1926.501							

APPLICABLE STANDARDS, GUIDES AND PUBLICATIONS

SUBJECT	USDL & USDI	NFPA	ANSI	API	MCA		NSC	MISC
(CON'T) GUARDRAILS								
STANDARD SPECIFICATIONS	1926.500							
STEEL ERECTION	1926.752							
TOEBOARDS	1910.24							
WEATHER DECKS	1918.36							
WELDING & CUTTING	1915.34							
(MULTIPLE REFERENCES)	1916.34							
(MULTIPLE REFERENCES)	1917.34							
GUILLOTINE CUTTERS								
POWER							ISDS298	
GUYS, GUYING, REINFORCING STEEL	1926.700							
BARGE	1918.37							
DECK LOADS	1918.41							
PLACEMENT	1918.54							
PREVENTERS	1918.54							
SECURING	1918.54							
TRUNK BULKHEADS	1918.41							

APPLICABLE STANDARDS, GUIDES AND PUBLICATIONS

SUBJECT	USDL & USDI	NFPA	ANSI	API	MCA		NSC	MISC
HAFNIUM POWDER					SD-92			
HALON 1211								
EXTINGUISHING SYSTEM		12B						
HALON 1301								
EXTINGUISHING SYSTEM		12A						
HAND								
JOINTERS	1910.214							
LAMPS	1926.402							
SAW, ELECTRIC							ISDS344	
SIGNALS	1926.551							
SOLDERING & BRAZING							ISDS445A	
TOOLS	1910.242							
(MULTIPLE REFERENCES)	1926.300		A10.3					
TRUCK	1910.263							
TRUCK, POWERED							ISDS317	
HAND TOOLS	1915.73							
(MULTIPLE REFERENCES)	1916.73							
(MULTIPLE REFERENCES)	1917.73							
HAND TOOLS, POWERED								
AIR POWERED HAND TOOL							ISDS392	
CHURN DRILL, BLAST HOLE DRILLING							ISDS200	
ELECTRIC HAND SAW							ISDS344	
JACK LEG DRILL							ISDS504	
PORTABLE GRINDER							ISDS583	
PORTABLE POWER CHAIN SAW							ISDS320	
PORTABLE REAMER-DRILL							ISDS497A	
POWDER ACTUATED TOOLS			A10.3				ISDS236	
POWER LAWN MOWERS							ISDS464	
RADIAL SAWS							ISDS353	
HANDHOLDS								
MANLIFTS	1910.68							
HANDLING MATERIALS	1910.176							
CLEARANCE LIMITS	1910.176							
COMPRESSED GASES	1910.107							
DRAINAGE	1910.176							
FLAMMABLE LIQUID	1910.107							
GUARDING	1910.176							
HOUSEKEEPING	1910.176							
LIQUID FUELS	1910.178							
MECHANICAL EQUIPMENT	1910.176							
ROLLING RAILROAD CARS	1910.176							
STORAGE	1910.176							

APPLICABLE STANDARDS, GUIDES AND PUBLICATIONS

SUBJECT	USDL & USDI	NFPA	ANSI	API	MCA		NSC	MISC
HANDRAILS (SEE GUARDRAILS)								
HARDHATS (SEE HEAD PROTECTION)								
HATCH	1910.23							
HATCH COVERING								
DAMAGED OR ILL FITTING	1918.31							
HINGED OR FOLDING	1918.43							
HYDRAULIC	1918.41							
INTERMEDIATE DECKS	1918.31							
MECHANICAL	1918.41							
MISSING	1918.31							
PARTIALLY OPENED	1918.43							
PLACEMENT	1918.31							
PONTOONS	1918.43							
REPAIRS	1918.31							
SECURING	1918.43							
STOWAGE	1918.43							
WEATHER DECKS	1918.43							
HATCH PROTECTION	1918.35							
HATS	1926.951							
SAFETY			Z89.1					
HAULER-LOADER							ISDS576	
HAULING, HAULAGE								
EQUIPMENT	1926.602							
HAZARD								ISO R408
COLOR CODING			Z53.1					
GAS ON SHIPS, CONTROL OF		306	Z270.1					
HAZARDOUS AREA								
ELECTRICAL STANDARDS					SG-19			
HAZARDOUS MATERIALS								
ACETYLENE	1910.102							
ADDITIONAL DELAY IN EFFECTIVE DATE	1910.114							
COMPRESSED GASES	1910.101							
DEFINITION	1915.2							
(MULTIPLE REFERENCES)	1916.2							
(MULTIPLE REFERENCES)	1917.2							
(MULTIPLE REFERENCES)	1918.3							
DIP TANKS CONTG FLAMMABLE OR COMBUSTIBLE LIQUIDS	1910.108							
EPOXY PRODUCTS - CLASSIFYING & LABELING			K68.1					
EXPLOSIVES & BLASTING AGENTS	1910.109							
FIRE CODE		49						

SUBJECT	USDL & USDI	NFPA	ANSI	API	MCA		NSC	MISC
	APPLICABLE STANDARDS, GUIDES AND PUBLICATIONS							
(CON'T)HAZARDOUS MATERIALS								
FLAMMABLE & COMBUSTIBLE LIQUIDS	1910.106							
HANDLING	1918.86							
HYDROGEN	1910.103							
INSTRUCTIONS	1915.57							
(MULTIPLE REFERENCES)	1916.57							
(MULTIPLE REFERENCES)	1917.57							
LABELING					L-1			
MATERIAL SAFETY DATA SHEET	1915.57							
(MULTIPLE REFERENCES)	1916.57							
(MULTIPLE REFERENCES)	1917.57							
NITROUS OXIDE	1910.105							
OXYGEN	1910.104							
PRECAUTIONS	1915.57							
(MULTIPLE REFERENCES)	1916.57							
(MULTIPLE REFERENCES)	1917.57							
SPILLAGE OR LEAKAGE	1918.86							
SPRAY FINISHING	1910.107							
STANDARDS ORGANIZATIONS	1910.116							
STANDARDS SOURCES	1910.115							
STORAGE & HANDLING - ANHYDROUS AMMONIA	1910.111							
STORAGE & HANDLING - LIQUEFIED PETROLEUM	1910.110							
VAPORIZING LIQUID FIRE EXTINGUISHING AGENTS		182M						
HAZARDOUS WASTES								
DISPOSAL					SG-9			
HEAD PROTECTION	1910.135							
(MULTIPLE REFERENCES)	1915.83							
(MULTIPLE REFERENCES)	1916.83							
(MULTIPLE REFERENCES)	1917.83							
(MULTIPLE REFERENCES)	1918.105							
(MULTIPLE REFERENCES)	1926.100							
FOR VEHICULAR USER			Z90.1					
HELMET FOR ELECTRICAL WORKERS			Z89.2					
SAFETY STANDARD			Z89.1					
HEADROOM, EXITS	1910.37							
HEARING								
PROTECTION	1910.95							
(MULTIPLE REFERENCES)	1926.101							
HEAT TREATING								
CARGO GEAR	1919.16							
EXEMPTIONS	1919.17							
GEAR, LOOSE	1919.80							
NITRATE-NITRITE SALT BATH							ISDS270	
OTHER THAN IRON	1919.16							
(MULTIPLE REFERENCES)	1919.36							

APPLICABLE STANDARDS, GUIDES AND PUBLICATIONS

SUBJECT	USDL & USDI	NFPA	ANSI	API	MCA		NSC	MISC
(CON'T)HEAT TREATING								
SUPERVISION	1919.10							
WROUGHT IRON	1919.36							
HEATERS, HEATING, TEMPORARY DEVICES	1926.154							
HEATING EQUIPMENT	1910.106							
CLEARANCE		89M						
ELECTRIC DRY BATH HEATER			C33.75					UL 875
ELECTRIC HEATER FOR HAZARDOUS LOCATION			C33.48					UL 823
ELECTRIC HEATING APPLIANCES			C33.11					UL 499
ELECTRIC HEATING PAD			C33.7					UL 130
ELECTRIC SPACE HEATING			C33.12					UL 573
ELECTRIC STORAGE-TANK WATER HEATER			C33.87					UL 174
GAS FIRED FORCED AIR HEATER			Z83.5					
HEAT, LIGHTS, INDOOR STORAGE	1910.177							
HEATERS, HEATING, TEMPORARY DEVICES	1926.154							
LIQUID FUELS	1910.177							
RADIANT HEAT CONTROL							ISDS381A	
RADIO FREQUENCY HEATING							ISDS319	
RESIDENTIAL TYPE		90B						
VENTING GUIDE		204						
HEAVY OIL								
STORAGE/LOADING SAFETY GUIDE				PSD-2205				
HELICOPTERS	1926.551							
(MULTIPLE REFERENCES)	1926.958	418						
APPROACH DISTANCE	1926.551							
APPROACHING	1926.551							
BRIEFINGS	1926.551							
CARGO HOOKS	1926.551							
CLOTHING	1926.551							
COMMUNICATIONS	1926.551							
FIRE	1926.551							
GROUND LINES	1926.551							
GROUND PERSONNEL	1926.551							
HOOKING & UNHOOKING LOADS	1926.551							
HOUSEKEEPING	1926.551							
LOOSE GEAR	1926.551							
OPERATOR RESPONSIBILITY	1926.551							
PROTECTIVE EQUIP	1926.551							
REGULATIONS	1926.551							
SIGNAL SYSTEMS	1926.551							
SLINGS	1926.551							
STATIC CHARGES	1926.551							
VISIBILITY	1926.551							
WEIGHT LIMITATIONS	1926.551							

APPLICABLE STANDARDS, GUIDES AND PUBLICATIONS

SUBJECT	USDL & USDI	NFPA	ANSI	API	MCA		NSC	MISC
HELMETS (SEE HEAD PROTECTION)								
HEPTANE								HG
HEXACHLOROCYCLOHEXANE								HG
HEXANE								HG
2-HEXANONE								HG
HEXONE (SEE METHYL ISOBUTYL KETONE)								
HIGH PRESSURE								
CYLINDERS	1910.166							
WATER BLASTING (HYDROBLASTING)							ISDS633	
HIGH VOLTAGE LINES	1926.550							
CIRCUIT, UNDERGROUND	75.800							
HIGHWAY								
ACCIDENT - CLASSIFICATION MANUAL			D16.1					
ACCIDENT - RECORDING & MEASURING			D15.1					
(MULTIPLE REFERENCES)			D15.2					
GLAZING MATERIAL FOR MOTOR VEHICLE			Z26.1					
LIGHTING			D12.1					IES RPB
RAILROAD GRADE CROSSING			D8.1					AAR BULL 6
TRAFFIC CONTROL - ADJUSTABLE FACE SIGNAL			D10.1					ITE TR1
TRAFFIC CONTROL - PRE-TIMED, FIXED SIGNAL			D11.1					ITE TR2
TRAFFIC CONTROL - TRAFFIC ACTUATED			D13.1					ITE TR3
TRAFFIC CONTROL MANUAL			D6.1					
HOGS & CHIPPERS							ISDS602	
HOISTS, HOISTING								
AUTOMOTIVE EQUIPMENT							ISDS437	
BASE-MOUNTED DRUM	1926.553							
CERTIFICATION, CRANES & DERRICKS	1917.65							
CONSTRUCTION MATERIAL							ISDS511A	
CRANE WITH ELECTROMAGNETS							ISDS359A	
CRANES & DERRICKS, CERTIFICATION	1915.65							
(MULTIPLE REFERENCES)	1916.65							
DRUM - BASE MOUNTED			B30.7					
EQUALIZER	1910.179							
EQUIPMENT - GENERAL	75.1400							
LIMIT SWITCHES	1910.179							
MARINE RAILWAYS	1915.65							
(MULTIPLE REFERENCES)	1916.65							
(MULTIPLE REFERENCES)	1917.65							
MATERIAL - SAFETY STANDARD			A10.5					
MATERIALS	1926.552							

APPLICABLE STANDARDS, GUIDES AND PUBLICATIONS

SUBJECT	USDL & USDI	NFPA	ANSI	API	MCA		NSC	MISC
(CON'T)HOISTS, HOISTING								
MOVING PARTS	1915.65							
(MULTIPLE REFERENCES)	1916.65							
(MULTIPLE REFERENCES)	1917.65							
OVERHEAD	1926.554		B30.16					
PERSONNEL (AND ELEVATORS)	1926.552							
ROPE	1910.179							
SAFE WORKING LOADS	1915.65							
(MULTIPLE REFERENCES)	1916.65							
(MULTIPLE REFERENCES)	1917.65							
SWING RADIUS	1915.65							
(MULTIPLE REFERENCES)	1916.65							
(MULTIPLE REFERENCES)	1917.65							
WORKMEN'S - SAFETY STANDARD			A10.4					
HOLES	1910.23							
HOOKS	1910.179							
(MULTIPLE REFERENCES)	1915.63							
(MULTIPLE REFERENCES)	1916.63							
(MULTIPLE REFERENCES)	1917.63							
HOOP, DRIVERS & TRUSSERS	1910.214							
FLARING & EXPANDING MACHINES	1910.214							
PUNCHING & COILING MACHINES	1910.214							
HOSES & CONNECTIONS								
ABRASIVE BLASTING	1915.23							
(MULTIPLE REFERENCES)	1916.23							
AIR LINE STEEL ERECTION	1926.752							
COUPLING, FLAMMABLE LIQUIDS	1910.106							
FIRE	1926.150							
FUEL GASES	1915.35							
(MULTIPLE REFERENCES)	1916.35							
(MULTIPLE REFERENCES)	1917.35							
MANIFOLDS	1926.350							
NOZZLES	1926.151							
OUTLETS, WATER, FIREFIGHTING	1910.158							
OXYGEN	1915.35							
(MULTIPLE REFERENCES)	1916.35							
(MULTIPLE REFERENCES)	1917.35							
OXYGEN, FUEL & GAS	1910.252							
PILE DRIVING	1926.603							
PNEUMATIC POWER TOOLS	1926.302							
STEAM	1915.53							
(MULTIPLE REFERENCES)	1916.53							
(MULTIPLE REFERENCES)	1917.53							

OCCUPATIONAL SAFETY, HEALTH & FIRE INDEX

SUBJECT	APPLICABLE STANDARDS, GUIDES AND PUBLICATIONS							
	USDL & USDI	NFPA	ANSI	API	MCA		NSC	MISC
HOSPITAL								
ELECTRICAL SYSTEMS		76A						
ELECTRICITY - SAFE USAGE		76BM						
FIRE EMERGENCY PREPAREDNESS		3M						
LABORATORY - OCCUPANCY STANDARD & PROCESS HAZARD		56C						
HOT TAPPING								
FLAMMABLE LIQUID PIPELINE/VESSEL/TANK				PSD-2201				
HOT WORK								
BELOW DECKS	1917.13							
CERTIFICATION	1915.13							
(MULTIPLE REFERENCES)	1917.13							
DRY CARGO HOLDS	1915.13							
ENGINE ROOM	1915.13							
(MULTIPLE REFERENCES)	1917.13							
OPEN	1917.14							
PERMIT (FLAME OR SPARK)							ISDS522	
TANK BOTTOM REPAIRS				PSD-2207				
WEATHER DECKS	1915.14							
HOUSEKEEPING	1910.141							
(MULTIPLE REFERENCES)	1918.91							
(MULTIPLE REFERENCES)	1926.25							
ACCESS	1915.51							
(MULTIPLE REFERENCES)	1916.51							
(MULTIPLE REFERENCES)	1917.51							
ASBESTOS	1910.93							
CHEMICAL INDUSTRY					SG-2			
CRANE CABS	1910.179							
FLAMMABLE LIQUID	1910.106							
FLAMMABLE SUBSTANCES	1915.51							
(MULTIPLE REFERENCES)	1916.51							
MATERIALS HANDLING & STORAGE	1910.176							
PASSAGEWAYS	1910.22							
REQUIREMENTS	1915.51							
(MULTIPLE REFERENCES)	1916.51							
(MULTIPLE REFERENCES)	1917.51							
SLIPPERY CONDITIONS	1915.51							
(MULTIPLE REFERENCES)	1916.51							
WALKING-WORKING SURFACES	1910.22							
HYDRAULIC								
JACKS	1926.305							
POWER TOOLS	1926.302							

APPLICABLE STANDARDS, GUIDES AND PUBLICATIONS

SUBJECT	USDL & USDI	NFPA	ANSI	API	MCA		NSC	MISC
HYDRAULIC FLUID							ISDS471	
FIRE-RESISTANT WATER-IN-OIL EMULSION							ISDS543	
HYDRAZINE					CC-7			
(MULTIPLE REFERENCES)					TEG			HG
HYDRAZINE/UDMH					CC-8			
(MULTIPLE REFERENCES)					CC-9			
(MULTIPLE REFERENCES)					TEG			
HYDRIDES							ISDS508A	
BORON							ISDS462	
METAL								
HYDROBLASTING							ISDS633	
HYDROCHLORIC ACID					CC-82			
(MULTIPLE REFERENCES)					CIC-51			
(MULTIPLE REFERENCES)					SD-39			
(MULTIPLE REFERENCES)					TEG			
TANK CAR UNLOADING					TC-2			
HYDROCYANIC ACID					SD-67			
HYDROFLUORIC ACID					CC-42			
(MULTIPLE REFERENCES)					CIC-52			
(MULTIPLE REFERENCES)					SD-25			
(MULTIPLE REFERENCES)					TEG		ISDS459	
HYDROGEN	1910.103							
FEDERAL SUPPLY CONTRACTS	50-204.68							
GASEOUS HYDROGEN SYSTEMS	1910.103							
GENERAL	1910.103							
LIQUEFIED HYDROGEN SYSTEMS	1910.103							
TESTING LOW PRESSURE CYLINDERS	1910.166							
HYDROGEN (LIQUID)					CC-12			
(MULTIPLE REFERENCES)					TEG			
HYDROGEN CHLORIDE					CIC-53			HG
HYDROGEN CYANIDE								HG
HYDROGEN FLUORIDE					CIC-54			HG
TOXIC EFFECT			Z37.28					

SUBJECT	USDL & USDI	NFPA	ANSI	API	MCA		NSC	MISC
	APPLICABLE STANDARDS, GUIDES AND PUBLICATIONS							
HYDROGEN PEROXIDE					CC			
(MULTIPLE REFERENCES)					SD-53			
(MULTIPLE REFERENCES)					TEG			HG
HYDROGEN PEROXIDE (HIGH STRENGTH)					CC-14			
(MULTIPLE REFERENCES)					SD-53			
(MULTIPLE REFERENCES)					TEG			
HYDROGEN PEROXIDE (NOT EXCEEDING 52%)					SD-53			
HYDROGEN SELENIDE								HG
HYDROGEN SULFIDE					SD-36		ISDS284	HG
EXPOSURE LEVEL	1910.93							
TOXIC EFFECT			Z37.2					
HYDROGEN SYSTEMS								
GASEOUS	1910.103	50A						
LIQUEFIED	1910.103							
MAINTENANCE	1910.103							
MARKINGS	1910.103							
SAFETY RELIEF DEVICES	1910.103							
TESTING	1910.103							
HYDROQUINONE								HG
HYDROSTATIC TESTS								
FIRE EXTINGUISHERS	1910.157							
LOW PRESSURE CYLINDERS	1910.166							
PIPING	1910.106							
STANDPIPE & HOSE SYSTEMS	1910.158							
HYPERBARIC FACILITIES								
GAS SYSTEM		56D						
HYPOBARIC FACILITIES								
GAS SYSTEM		56E						

OCCUPATIONAL SAFETY, HEALTH & FIRE INDEX

SUBJECT	APPLICABLE STANDARDS, GUIDES AND PUBLICATIONS							
	USDL & USDI	NFPA	ANSI	API	MCA		NSC	MISC
ICE	1918.91							
ICE REMOVAL, INDUSTRIAL							ISDS402	
IDENTIFICATION								
AGRICULTURAL MACHINERY - OPERATOR CONTROLS			B114.2					ASAE S304.3
(MULTIPLE REFERENCES)								SAE J389
AGRICULTURAL TRACTOR - OPERATOR CONTROLS			B114.2					ASAE S304.2
(MULTIPLE REFERENCES)								SAE J389
BADGES	1926.803							
BENZIDINE - REGULATED AREA REQUIREMENTS	1910.93							
BENZIDINE CONTAINER	1910.93							
COMPRESSED GASES				PSD-2206				
MATERIALS					SG-18			
PIPING SYSTEMS			A13.1					
VEHICLE - SLOW MOVING			B114.1					ASAE S276.2
(MULTIPLE REFERENCES)								SAE J943
IGNITION								
CABLE - SAFETY STANDARD			C33.23					UL 814
VAPORS BY FLASHLIGHT				PSD-2212				
IGNITION HAZARDS	1926.151							
IGNITION SOURCES	1910.108							
COMBUSTIBLE LIQUIDS	1910.106							
DISTANCES FROM HAZARDS	1910.107							
ELECTRIC WIRING	1910.107							
GROUNDING	1910.107							
SPRAY FINISHING	1910.107							
ILLUMINATION (SEE LIGHTING)								
IMPACT SENSITIVE MATERIAL								
MATERIALS HANDLING					SG-7			
STORAGE					SG-7			
IMPACT TOOL								
SAFE-ENDING							ISDS414	
INCIDENT REPORTING INVOLVING RADIATION	1910.96							
INCINERATOR								
RUBBISH HANDLING		82						
INDUSTRIAL DISPOSAL SYSTEMS	1910.143							

SUBJECT	USDL & USDI	NFPA	ANSI	API	MCA		NSC	MISC
	APPLICABLE STANDARDS, GUIDES AND PUBLICATIONS							
INDUSTRIAL HYGIENE (SEE ALSO FIRST AID &/OR MEDICAL)								
AIR QUALITY CRITERIA & OXIDANTS				AQM 69-5				
AIR QUALITY CRITERIA & PARTICULATES				AQM 69-2				
BERYLLIUM							ISDS562	
CHEMICAL BURNS							ISDS523	
CHEMICAL SAFETY REFERENCES							ISDS486A	
CLEANING MACHINERY & ELECTRIC MOTORS							ISDS285	
COMBUSTIBLE METALS, FIRE PROTECTION							ISDS567	
DEGREASING, LIQUID							ISDS537	
DUSTS, FUMES, MISTS							ISDS531	
EPOXY RESIN SYSTEM							ISDS533A	
EXHAUST SYSTEM (LOCAL) PERFORMANCE							ISDS428	
EXHAUST VENTILATION SYSTEM, TEST INSTRUMENTS							ISDS431	
GAS-SHIELDED ARC & PLASMA JET TORCH							ISDS552	
NOISE, JET ENGINE							ISDS580	
NUCLEAR ENERGY, SOURCES OF INFORMATION							ISDS446	
PAINT SPRAYING & DETEARING ELECTROSTATIC							ISDS468	
PAINTING, AIRLESS SPRAY							ISDS548	
PAPER BAG & SACK MANUFACTURE			Z259.1					
PETROLEUM ASPHALT				TB-7103				
PIPELINE RADIOGRAPHY							ISDS600	
PROTECTIVE EQUIP - ELECTRICAL WORKERS, FLEX INSUL							ISDS598	
PROTECTIVE EQUIP - RESPIRATORY							ISDS444A	
RADAR HAZARDS							ISDS481	
RADIANT HEAT CONTROL							ISDS381A	
SKIN DISEASES, INDUSTRIAL							ISDS510	
SUBSURFACE STRUCTURES & SEWERS, ATMOSPHERE							ISDS550	
X-RAY IN INDUSTRY							ISDS475	
INDUSTRIAL PLANTS								
COMBUSTIBLE & FLAMMABLE LIQUIDS	1910.106							
DRAINAGE	1910.106							
DUST EXPLOSION		63	Z12.21					
ELECTRIC WIRING	1910.106							
FIRE PROTECTION	1910.106							
LOADING	1910.106							
MAINTENANCE	1910.106							
INDUSTRIAL TRUCKS, POWERED	1910.178							
(MULTIPLE REFERENCES)	1926.602							
CHANGING & CHARGING BATTERIES	1910.178							
CONTROL OF NOXIOUS GASES & FUMES	1910.178							
CONVERTED INDUSTRIAL TRUCKS	1910.178							
DESIGNATED LOCATIONS	1910.178							
DESIGNATION	1910.178							
DOCKBOARDS, BRIDGE PLATES	1910.178							
FUEL HANDLING & STORAGE	1910.178							
GENERAL REQUIREMENTS	1910.178							
LIGHTING FOR OPERATING AREAS	1910.178							
LOADING	1910.178							

APPLICABLE STANDARDS, GUIDES AND PUBLICATIONS

SUBJECT	USDL & USDI	NFPA	ANSI	API	MCA		NSC	MISC
(CON'T) INDUSTRIAL TRUCKS, POWERED								
MAINTENANCE	1910.178							
NOXIOUS GASES & FUMES	1910.178							
OPERATION	1910.178							
OPERATOR TRAINING	1910.178							
SAFETY GUARDS	1910.178							
TRAVELING	1910.178							
TRUCK OPERATIONS	1910.178							
INERT GAS METAL-ARC WELDING	1926.353							
INJECTION MOLDING							ISDS454A	
INJURY & ILLNESS								
RECORDING & MEASURING	1904.1		Z16.1					
RECORDING & MEASURING STATISTICS			Z108.1					
RECORDING - NATURE & OCCURRENCE	1904.2		Z16.2					
REPORTING	1904.5							
SULFUR DIOXIDE - FACTS & OPINIONS				AQM 69-10				
INSECT CONTROL, LABOR CAMPS	1910.142							
INSPECTION	1926.955							
ALUMINUM CYLINDER	1910.166							
CHAINS	1918.64							
COMPRESSED GAS CYLINDERS	1910.101							
COOPERAGE MACHINERY	1910.214							
CRANES & DERRICKS	1926.550							
CRANES, LOCOMOTIVE, CRAWLER & TRUCK	1910.180							
DERRICKS	1910.181							
ELECTRICAL EQUIP - NUCLEAR POWER GENERATING STA			N45.2.4					IEEE 336
ELECTRICAL SYSTEMS - MODEL STATE LAW		70L						
EXCAVATIONS	1926.650							
(MULTIPLE REFERENCES)	1926.651							
FIRE PROTECTION EQUIPMENT					SG-13			
HOISTS & ELEVATORS	1926.552							
HOOKS	1918.66							
INSTRUMENTATION - NUCLEAR POWER GENERATING STA			N45.2.4					IEEE 336
MOTOR VEHICLE			D7.1					
MOTORCYLE			D7.5					
OVERHEAD GANTRY CRANES	1910.179							
PLANT SAFETY COMMITTEE					SG-20			
POWER PRESS	1910.217						ISDS603	
RESPIRATORY PROTECTION EQUIPMENT	1918.102							
ROPE, OVERHEAD GANTRY CRANE	1910.179							
SCAFFOLDS	1926.451							
SCHOOL BUS			D7.4					
SHORING	1926.700							
STEVEDORE GEAR	1918.61							
SUPERVISION	1919.10							

APPLICABLE STANDARDS, GUIDES AND PUBLICATIONS

SUBJECT	USDL & USDI	NFPA	ANSI	API	MCA		NSC	MISC
(CON'T) INSPECTION								
VISUAL	1919.26							
WOODWORKING MACHINERY	1910.213							
INSPECTION STATION								
MOTOR VEHICLE			D7.2					
(MULTIPLE REFERENCES)			D7.3					
INSTALLATION								
COMMUNICATION LINES			C2.3					
ELECTRIC SUPPLY & COMMUNICATION LINES			C2.2					NBS H81
ELECTRIC SUPPLY STATION			C2.1					
ELECTRIC SUPPLY, UNDERGROUND			C2.3					
ELECTRICAL EQUIP - NUCLEAR POWER GENERATING STA			N45.2.4					IEEE 336
FIRE EXTINGUISHERS, PORTABLE		10						
INSTRUMENTATION - NUCLEAR POWER GENERATING STA			N45.2.4					IEEE 336
INSULATORS	1910.107							
INTERNAL COMBUSTION ENGINES	1910.110							
(MULTIPLE REFERENCES)	1915.76							
(MULTIPLE REFERENCES)	1916.76							
(MULTIPLE REFERENCES)	1917.76							
IODINE							ISDS457A	HG
IONIZING RADIATION (SEE ALSO RADIATION)	1910.96							
(MULTIPLE REFERENCES)	1926.53							
ATOMIC ENERGY COMMISSION LICENSEE	1910.96							
CAUTION SIGNS, LABELS & SIGNALS	1910.96							
DEFINITION	1910.96							
DISCLOSURE TO FORMER EMPLOYEE OF EMPLOYEE RECORD	1910.96							
EXCEPTIONS FOR RADIOACTIVE MTRLS PACKED FOR SHPM	1910.96							
EXPOSURE TO AIRBORNE RADIOACTIVE MATERIALS	1910.96							
EXPOSURE TO RADIATION IN RESTRICTED AREAS	1910.96							
FISSIONABLE MATERIAL	1915.37							
(MULTIPLE REFERENCES)	1916.37							
IMMEDIATE EVACUATION WARNING SIGNALS	1910.96							
INSTRUCTION OF PERSONNEL POSTINGS	1910.96							
NOTIFICATION OF INCIDENTS	1910.96							
PRECAUTIONARY PROCEDURES & PERSONAL MONITORING	1910.96							
RADAR OR RADIO NEARBY	1915.55							
(MULTIPLE REFERENCES)	1916.55							
RADIATION STANDARDS FOR MINING	1910.96							
RECORDS	1910.96							
REPORTS OF OVEREXPOSURE & EXCESSIVE LEVELS	1910.96							
STORAGE OF RADIOACTIVE MATERIALS	1910.96							
WASTE DISPOSAL	1910.96							

SUBJECT	APPLICABLE STANDARDS, GUIDES AND PUBLICATIONS							
	USDL & USDI	NFPA	ANSI	API	MCA		NSC	MISC
ISOBUTANOL (SEE ISOBUTYL ALCOHOL)								
ISOBUTYL ACRYLATE					CIC-55			
ISOBUTYRALDEHYDE					CIC-56			
ISOCYANATES (TDI & MDI)							ISDS489A	
ISOPRENE					CIC-57			
ISOPROPANOL (SEE ISOPROPYL ALCOHOL)								
ISOPROPYL ALCOHOL (MULTIPLE REFERENCES) (MULTIPLE REFERENCES)					CC-71 SD-98 TEG			
ISOPROPYLAMINE (MULTIPLE REFERENCES) (MULTIPLE REFERENCES)					CC-56 SD-72 TEG			
ISOPROPYL ETHER (MULTIPLE REFERENCES)					CC-58 TEG			

APPLICABLE STANDARDS, GUIDES AND PUBLICATIONS

SUBJECT	USDL & USDI	NFPA	ANSI	API	MCA		NSC	MISC
JACK LEG DRILL							ISDS504	
JACKS	1926.305							
LEVER & RATCHET JACKS	1926.305							
LOADING & OPERATION	1910.244							
SAFETY CODE			B30.1					
TERMS	1910.241							
TRENCHING	1926.652							
JACOB'S LADDERS	1926.605							
JET ENGINE								
NOISE							ISDS580	
JOINTERS	1910.213							
BLADES	1910.213							
HAND	1910.214							
HEADING	1910.214							
STAVE	1910.214							
WHEELS	1910.214							

	APPLICABLE STANDARDS, GUIDES AND PUBLICATIONS							
SUBJECT	USDL & USDI	NFPA	ANSI	API	MCA		NSC	MISC
KEROSINE				TR				
KEYS, PROJECTING	1910.219							
KICK (FOOT) PRESS							ISDS363	
KITCHENS, LABOR CAMPS								
CONSTRUCTION & OPERATION	1910.142							
LIGHTING	1910.142							
REFUSE DISPOSAL	1910.142							
KNIVES							ISDS369	
KNOTS	1910.21							
WIRE ROPE	1918.63							

SUBJECT	USDL & USDI	NFPA	ANSI	API	MCA		NSC	MISC
			APPLICABLE STANDARDS, GUIDES AND PUBLICATIONS					
LABELING								
COLOR CODING PHYSICAL HAZARDS			Z53.1					ISO R408
COMPRESSED GAS CONTAINERS - PORTABLE			Z48.1					CGA C-4
(MULTIPLE REFERENCES)								ISO R443
(MULTIPLE REFERENCES)								ISO R448
EPOXY PRODUCTS BY HAZARD			K68.1					
HAZARDOUS CHEMICALS					L-1			
PRECAUTIONARY				TB-2511				
LABOR CAMPS	1910.142							
LADDERS	1926.951							
ACCESS, CARGO SPACES	1915.46							
(MULTIPLE REFERENCES)	1916.46							
(MULTIPLE REFERENCES)	1917.46							
ACCESS, STAGING	1915.41							
(MULTIPLE REFERENCES)	1916.41							
(MULTIPLE REFERENCES)	1917.41							
ACCESS, VESSELS	1915.44							
(MULTIPLE REFERENCES)	1916.44							
(MULTIPLE REFERENCES)	1917.44							
CHICKEN	1926.451							
CONDUCTIVE	1926.951							
CRANES & DERRICKS	1926.550							
FIRE DEPARTMENT		193						
FIXED			A14.3				ISDS606	
HOOK	1926.951							
JACOB'S	1926.605							
JACOB'S LADDERS	1915.44							
(MULTIPLE REFERENCES)	1916.44							
(MULTIPLE REFERENCES)	1917.44							
JOB-MADE			A14.4				ISDS568	
LADDER JACK SCAFFOLDS	1926.451							
LADDER-TYPE PLATFORMS	1926.451							
MARKINGS	1926.951							
MASON'S	1910.25							
MINES, FOR - CONSTRUCTION & MAINTENANCE			M12.1					
MOBILE - MANUALLY PROPELLED			A92.1					
MOBILE STANDS	1910.29							
NONCONDUCTIVE	1926.951							
PILE DRIVING	1926.603							
PLATFORM GUARDS	1910.23							
PORTABLE	1926.951							
PORTABLE - METAL			A14.2					
PORTABLE - WOOD			A14.1					
PORTABLE WOOD	1917.42							
PORTABLE WOOD, CONSTRUCTION OF	1915.42							
(MULTIPLE REFERENCES)	1916.42							
REQUIREMENTS	1915.42							
(MULTIPLE REFERENCES)	1916.42							
(MULTIPLE REFERENCES)	1917.42							

SUBJECT	USDL & USDI	NFPA	ANSI	API	MCA		NSC	MISC
	APPLICABLE STANDARDS, GUIDES AND PUBLICATIONS							
(CON'T)LADDERS								
SECURING	1926.951							
WOOD & EXTENSION TRESTLE LADDERS	1915.41							
(MULTIPLE REFERENCES)	1916.41							
LADDERS & FOOTWALKS	1910.179							
LADDERS, CARGO SPACES								
ALTERNATIVES	1918.25							
GENERAL	1918.25							
RUNG CLEARANCE	1918.25							
UNSAFE	1918.25							
LADDERS, FIXED	1910.27							
CAGES OR WELLS	1910.27							
CLEARANCE	1910.27							
CLEARANCE CLIMBING SIDE	1910.27							
CLEARANCE HATCH COVERS	1910.27							
CLEARANCE IN BACK	1910.27							
CLEARANCE OF GRAB BAR	1910.27							
CLEARANCE STEP-ACROSS DISTANCE	1910.27							
CLEARANCE WITH CAGES	1910.27							
CLEARANCE WITHOUT CAGES	1910.27							
DESIGN REQUIREMENTS	1910.27							
EXTENSIONS	1910.27							
LANDING PLATFORMS	1910.27							
MAINTENANCE	1910.27							
PITCH	1910.27							
REQUIREMENTS	1910.27							
SPECIFIC FEATURES	1910.27							
LADDERS, JACOBS	1926.605							
ACCEPTABLE	1918.22							
BARGE OR TOWBOAT	1918.23							
IN LIEU OF FIXED HOLD LADDERS	1918.25							
IN LIEU OF GANGWAYS	1918.11							
SUSPENSION	1918.22							
LADDERS, METAL PORTABLE	1910.26							
CARE	1910.26							
PLATFORMS	1910.26							
REQUIREMENTS	1910.26							
STEPLADDERS	1910.26							
STRAIGHT EXTENSION	1910.26							
TESTING	1910.26							
TRESTLES & EXTENSION	1910.26							

SUBJECT	USDL & USDI	NFPA	ANSI	API	MCA		NSC	MISC
	APPLICABLE STANDARDS, GUIDES AND PUBLICATIONS							
LADDERS, STRAIGHT PORTABLE								
BARGE OR TOWBOAT	1918.24							
IN LIEU OF FIXED HOLD LADDERS	1918.25							
IN LIEU OF GANGWAYS	1918.11							
SECURED	1918.25							
LADDERS, TO WATER	1918.96							
LADDERS, WOOD PORTABLE	1910.25							
CARE	1910.25							
CONSTRUCTION REQUIREMENTS	1910.25							
MATERIAL REQUIREMENTS	1910.25							
REQUIREMENTS	1910.25							
LADDERWAY GUARDING	1910.23							
LADINGS								
FROZEN, THAWING							ISDS233	
LAMPHOLDERS								
EDISON - BASE			C33.17					UL 496
LANDINGS	1910.68							
LANYARDS (SEE ALSO BODY BELTS,SAFETY STRAPS)	1926.104							
(MULTIPLE REFERENCES)	1926.556							
(MULTIPLE REFERENCES)	1926.959							
PILE DRIVING	1926.603							
LASERS								
EYE PROTECTION	1926.102							
NONIONIZING RADIATION	1926.54							
SAFE USE OF			Z136.1					
LATHES								
ENGINE							ISDS264	
PROFILE	1910.213							
WOOD	1910.213						ISDS253A	
LAUNDRY EQUIPMENT								
ELECTRIC HOME - LAUNDRY			C33.13					UL 560
SAFETY CODE			Z8.1					
LAUNDRY FACILITIES, LABOR CAMPS	1910.142							
LAUNDRY OPERATIONS	1910.264							
DEFINITION	1910.264							
DRYING MACHINES	1910.264							
EXTRACTORS	1910.264							
FINISHING MACHINES	1910.264							

SUBJECT	USDL & USDI	NFPA	ANSI	API	MCA		NSC	MISC
	APPLICABLE STANDARDS, GUIDES AND PUBLICATIONS							
(CON'T)LAUNDRY OPERATIONS								
GENERAL REQUIREMENTS	1910.264							
IRONERS	1910.264							
LIGHTING	1910.264							
MACHINERY GUARDING	1910.264							
MISCELLANEOUS EQUIPMENT	1910.264							
OPERATING RULES	1910.264							
POINT-OF-OPERATION GUARDS	1910.264							
SAFETY STANDARD	1910.264							
SEWING MACHINES	1910.264							
SHAKERS	1910.264							
STARCHING MACHINES	1910.264							
VENTILATION	1910.264							
WASHROOM MACHINES	1910.264							
LAVATORIES	1910.141							
LAWN MOWER								
POWER			B71.1				ISDS464	
POWER GUARDS	1910.243							
ROTARY	1910.243							
LEAD & ITS COMPOUNDS							ISDS443	HG
AIR QUALITY STANDARDS				AQM 69-11				
BURNING	1910.261							
TOXIC EFFECT			Z37.11	AQM 69-7				
(MULTIPLE REFERENCES)				MRR-7102				
WELDING	1910.252							
LEAD OXIDE					SD-64			
LEADED GASOLINE								
STORAGE TANK TOXIC EFFECT				PSD-2202				
LEAKS								
ELECTRIC CURRENT - APPLIANCES			C101.1					
UNDERGROUND FLAMMABLE LIQUID STORAGE TANK		329						
LENSES	1926.102							
LASERS	1926.102							
WELDING	1926.102							
LEVER & RATCHET JACKS	1926.305							
LIFE SAFETY CODE								
BUILDING CONSTRUCTION & MATERIALS		101	A9.1					

SUBJECT	USDL & USDI	NFPA	ANSI	API	MCA		NSC	MISC
	APPLICABLE STANDARDS, GUIDES AND PUBLICATIONS							
LIFEBOATS								
PERSONNEL	1915.56							
(MULTIPLE REFERENCES)	1916.56							
(MULTIPLE REFERENCES)	1917.56							
PRECAUTIONS	1915.56							
(MULTIPLE REFERENCES)	1916.56							
(MULTIPLE REFERENCES)	1917.56							
LIFEJACKETS								
REQUIREMENTS	1915.84							
(MULTIPLE REFERENCES)	1916.84							
(MULTIPLE REFERENCES)	1917.84							
SMALL BOATS	1915.47							
(MULTIPLE REFERENCES)	1916.47							
(MULTIPLE REFERENCES)	1917.47							
WORKING NEAR DECK EDGES	1915.43							
(MULTIPLE REFERENCES)	1916.43							
LIFELINES	1910.66							
(MULTIPLE REFERENCES)	1917.84							
(MULTIPLE REFERENCES)	1918.96							
(MULTIPLE REFERENCES)	1926.104							
MATERIALS STORAGE	1926.250							
REQUIREMENTS	1915.84							
(MULTIPLE REFERENCES)	1916.84							
SCAFFOLDING	1926.451							
WORKING MORE THAN 5 FT. ALOFT	1915.47							
(MULTIPLE REFERENCES)	1916.47							
WORKING ON SWINGING SCAFFOLDS	1915.41							
(MULTIPLE REFERENCES)	1916.41							
LIFERINGS	1915.84							
(MULTIPLE REFERENCES)	1916.84							
(MULTIPLE REFERENCES)	1917.84							
(MULTIPLE REFERENCES)	1918.96							
LIFESAVING EQUIPMENT	1926.100							
LOGGING	1910.265							
MARINE OPERATIONS	1910.28							
(MULTIPLE REFERENCES)	1926.605							
LIFT SLAB CONSTRUCTION	1926.305							
LIFTING EQUIPMENT	1926.602							
AUTOMOTIVE - CONSTRUCTION, CARE, USE			B153.1					

APPLICABLE STANDARDS, GUIDES AND PUBLICATIONS

SUBJECT	USDL & USDI	NFPA	ANSI	API	MCA		NSC	MISC
LIGHTING	1918.92							
(MULTIPLE REFERENCES)	1926.26							
(MULTIPLE REFERENCES)	1926.56							
ACCESS MEANS	1915.52							
(MULTIPLE REFERENCES)	1916.52							
(MULTIPLE REFERENCES)	1917.52							
CHRISTMAS - TREE			C33.25					UL 588
COLD WORK	1915.12							
(MULTIPLE REFERENCES)	1917.12							
COLD WORK - PORTABLE	1915.12							
(MULTIPLE REFERENCES)	1917.12							
CONTAINER AREAS	1910.110							
CRANES	1910.179							
DECORATIVE			C33.25					UL 588
EMERGENCY							ISDS248A	
EXIT & EXIT SIGNS	1910.37							
FIXTURES FOR HAZARDOUS LOCATION			C33.28					UL 844
GROUNDING	1915.52							
(MULTIPLE REFERENCES)	1916.52							
(MULTIPLE REFERENCES)	1917.52							
GROUNDING - PORTABLE	1915.52							
(MULTIPLE REFERENCES)	1916.52							
(MULTIPLE REFERENCES)	1917.52							
ILLUMINATION - PORTABLE	1915.52							
(MULTIPLE REFERENCES)	1916.52							
(MULTIPLE REFERENCES)	1917.52							
INDUSTRIAL			A11.1					IES RP7
INDUSTRIAL TRUCKS	1910.178							
LABOR CAMPS	1910.142							
LAUNDRIES	1910.264							
(MULTIPLE REFERENCES)	1910.264							
MACHINERY, BASEMENT AREAS	1910.219							
MANLIFTS	1910.68		A90.1					
OFFICE			A132.1					IES RP1
OPEN FLAMES PROHIBITED	1918.92							
OPEN-PIT MINE	55.17							
OPERATING AREAS, INDUSTRIAL TRUCKS	1910.178							
PAINTING	1915.24							
(MULTIPLE REFERENCES)	1916.24							
PORTABLE ELECTRIC - FOR HAZARDOUS LOCATION			C33.79					UL 781
PORTABLE EQUIPMENT	1926.151							
PORTABLE LIGHTS	1915.52							
(MULTIPLE REFERENCES)	1916.52							
(MULTIPLE REFERENCES)	1917.52							
PROTECTIVE			A85.1					IES RP10
PULP, PAPER & PAPERBOARD MILLS	1910.261							
ROADWAY			D12.1					IES RP8
SAND, GRAVEL, CRUSHED STONE OPERATION	56.17							
SCHOOL			A23.1					IES RP3
SPRAY BOOTHS	1910.107							

SUBJECT	USDL & USDI	NFPA	ANSI	API	MCA		NSC	MISC
	APPLICABLE STANDARDS, GUIDES AND PUBLICATIONS							
(CON'T)LIGHTING								
STORAGE AREAS	1910.177							
(MULTIPLE REFERENCES)	1910.178							
TEMPORARY	1915.52							
(MULTIPLE REFERENCES)	1916.52							
(MULTIPLE REFERENCES)	1917.52							
(MULTIPLE REFERENCES)	1926.401							
UNDERGROUND MINE	56.17							
LIGHTNING								
PETROLEUM PRODUCTS FIRE PROTECTION				RP-2003				
PROTECTION CODE		78	C5.1					
LIME							ISDS241A	
LIMIT CONTROL								
SAFETY STANDARD			B136.1					UL 353
LINEMEN CLIMBERS								
CARE & MAINTENANCE							ISDS620	
LINES, UNDERGROUND (SEE UNDERGROUND)								
LIQUEFIED HYDROGEN SYSTEMS								
ELECTRIC WIRING	1910.103							
GROUNDING	1910.103							
MAINTENANCE	1910.103							
MARKINGS	1910.103							
SAFETY RELIEF DEVICES	1910.103							
TESTING	1910.103							
LIQUEFIED NATURAL GAS								
STORAGE & HANDLING		59A	Z225.1					
LIQUEFIED PETROLEUM GAS (LPG)	1926.153							
AT UTILITY GAS PLANTS		59						
BASIC RULES	1910.110							
CONTAINER STORAGE	1910.110							
CYLINDER SYSTEMS	1910.110							
ELECTRIC WIRING	1910.110							
FIRE PROTECTION	1910.110							
INDUSTRIAL TRUCKS							ISDS479A	
LIQUID TRANSFER	1910.110							
LOADING	1910.110							
LPG EQUIPMENT	1926.153							
LUNCH ROOMS	1910.141							
MARKINGS	1910.110							
MOTOR FUELS	1910.110							
PIPE CONNECTORS			B148.1					UL 567
PIPELINE REPAIR GUIDE				PSD-2200				

SUBJECT	USDL & USDI	NFPA	ANSI	API	MCA		NSC	MISC
	APPLICABLE STANDARDS, GUIDES AND PUBLICATIONS							
(CON'T)LIQUEFIED PETROLEUM GAS (LPG)								
PRESSURE REGULATORS - SAFETY STANDARDS			Z217.1					UL 144
SAFETY RELIEF DEVICES	1910.110							
SERVICE STATIONS	1910.110							
STORAGE & HANDLING	1910.110	58	Z106.1					
SYSTEMS IN COMMERCIAL VEHICLES	1910.110							
LIQUIDS, FLAMMABLE & COMBUSTIBLE	1926.152							
PROPERTIES		325M						
STORAGE	1910.106							
LIQUIDS, HANDLING FOR DIP TANKS	1910.108							
LIQUID CHEMICALS								
SAMPLING, TANK CAR					SG-16			
SAMPLING, TANK TRUCK					SG-16			
LIQUID FUELS								
HEATING SYSTEMS	1910.177							
STORAGE & HANDLING	1910.178							
LIQUID HEATERS, SPRAY	1910.107							
LIQUID PETROLEUM								
PIPING			B31.4					
LIQUID TRANSFER								
ANHYDROUS AMMONIA	1910.111							
FLAMMABLE LIQUID	1910.107							
LPG	1910.110							
LITHARGE (SEE LEAD OXIDE)								
LITHIUM				AQM 69-20			ISDS566	
LITHIUM HYDRIDE								HG
LIVE LINE								
AERIAL LIFTS	1926.955							
ARM CURRENT TESTS	1926.955							
BARE-HAND WORK	1926.955							
BOOM	1926.955							
BUCKET LINERS, CONDUCTIVE	1926.955							
BUCKETS, CONDUCTIVE	1926.955							
CIRCUIT INTERRUPTING DEVICES	1926.955							
ELECTRICAL STORMS	1926.955							
EMPLOYEE TRAINING	1926.955							
GROUND LEVEL CONTROL	1926.955							
HAND LINES	1926.955							
MINIMUM DISTANCES	1926.955							

SUBJECT	USDL & USDI	NFPA	ANSI	API	MCA		NSC	MISC
	APPLICABLE STANDARDS, GUIDES AND PUBLICATIONS							
(CON'T)LIVE LINE								
TOOLS							ISDS498	
LIVESTOCK								
FLUOROSIS				AQM 69-4				
LOAD ATTACHING	1926.953							
LOAD HANDLING								
CRANES	1910.180							
(MULTIPLE REFERENCES)	1926.550							
DERRICKS	1910.181							
(MULTIPLE REFERENCES)	1926.550							
HOISTS & ELEVATORS	1926.552							
HOOKING & UNHOOKING	1926.551							
OVERHEAD GANTRY CRANES	1910.179							
ROCKER TYPE OPERATION							ISDS209A	
STORAGE	1926.250							
LOAD RATINGS								
CRANES, LOCOMOTIVE, CRAWLER & TRUCK	1910.180							
DERRICKS	1910.181							
MARKINGS	1910.66							
(MULTIPLE REFERENCES)	1910.178							
(MULTIPLE REFERENCES)	1910.180							
(MULTIPLE REFERENCES)	1910.181							
MAXIMUM RATED LOADS	1926.451							
RATINGS	1926.400							
LOADING								
ASPHALT SAFETY GUIDE				PSD-2205				
BULK PLANTS	1910.106							
EXPLOSIVES	1910.109							
HEAVY OIL SAFETY GUIDE				PSD-2205				
INDUSTRIAL PLANTS	1910.106							
INDUSTRIAL TRUCKS	1910.178							
LPG	1910.110							
PROCESSING PLANTS	1910.106							
SCAFFOLDS	1910.29							
LOCKS								
MAN AND EMERGENCY	1926.803							
MEDICAL	1926.803							
LOG HANDLING (SEE ALSO SAWMILLS)	1910.265							

APPLICABLE STANDARDS, GUIDES AND PUBLICATIONS

SUBJECT	USDL & USDI	NFPA	ANSI	API	MCA		NSC	MISC
LONGSHORING								
SAFETY REQUIREMENTS			MH9.1					
LOOMS	1910.262							
LOW PRESSURE CYLINDER INSPECTION	1910.166							
EXTERIOR	1910.166							
FIRE DAMAGE	1910.166							
HYDROSTATIC TESTS	1910.166							
INTERIOR	1910.166							
PREPARATION	1910.166							
REJECTION	1910.166							
LOW PRESSURE TANKS	1910.106							
LPG (SEE LIQUEFIED PETROLEUM GAS (LPG))								
LUMBER HANDLING	1910.265							
LUMINESCENT MATERIAL							ISDS245	
LUNCH ROOMS	1910.141							

APPLICABLE STANDARDS, GUIDES AND PUBLICATIONS

SUBJECT	USDL & USDI	NFPA	ANSI	API	MCA		NSC	MISC
MACHINERY & MACHINE GUARDING								
ABRASIVE WHEELS	1910.215							
CLEANING							ISDS285A	
COOPERAGE	1910.214							
DEFINITION	1910.211							
FEDERAL SUPPLY CONTRACTS	50-204.5							
FORGING MACHINES	1910.218							
GENERAL REQUIREMENTS FOR ALL MACHINES	1910.212							
MECHANICAL POWER PRESSES	1910.217							
MECHANICAL POWER TRANSMISSION APPARATUS	1910.219							
MILLS & CALENDERS IN THE RUBBER & PLASTIC IND	1910.216							
OPERATOR	1918.97							
WOODWORKING MACHINERY REQUIREMENTS	1910.213							
MACHINERY, ABRASIVE WHEELS	1910.215							
COOPERAGE	1910.214							
COOPERAGE MAINTENANCE	1910.214							
MACHINES, SANDING	1910.213							
MAGAZINES	1910.109							
MAGNESIUM							ISDS426	HG
HANDLING		48						
STORAGE		48						
MAGNESIUM POWDER PLANT								
DUST EXPLOSION			Z12.15					
MAINTENANCE								
AIRCRAFT CABIN CLEANING		410F						
AIRCRAFT CLEANING, PAINTING, PAINT REMOVAL		410D						
AIRCRAFT ELECTRICAL MAINTENANCE		410A						
AIRCRAFT FUEL RAMP DRAINAGE		415						
AIRCRAFT FUEL SYSTEM		410C	Z119.2					
AIRCRAFT OXYGEN		410B	Z227.1					
AUTOMOTIVE LIFT			B153.1					
BUILDING REPAIR, EXTERIOR	1910.66							
BULK OXYGEN SYSTEMS	1910.104							
COMMUNICATION LINES			C2.3					
CRANES, LOCOMOTIVE, CRAWLER & TRUCK	1910.180							
CRANES, OVERHEAD & GANTRY	1910.179							
DERRICKS	1910.181							
DIE CASTING MACHINE			B152.1					
DIP TANKS	1910.108	34						
EGRESS	1910.36							
ELECTRIC SUPPLY & COMMUNICATION LINES			C2.2					NBS H81
ELECTRIC SUPPLY STATION			C2.1					
ELECTRIC SUPPLY, UNDERGROUND			C2.3					
EXIT	1910.37							

SUBJECT	USDL & USDI	NFPA	ANSI	API	MCA		NSC	MISC
	APPLICABLE STANDARDS, GUIDES AND PUBLICATIONS							
(CON'T)MAINTENANCE								
FIRE ALARM SYSTEM	1910.163							
FIRE EXTINGUISHERS, CARBON DIOXIDE	1910.161							
FIRE EXTINGUISHERS, DRY CHEMICAL	1910.160							
FIRE EXTINGUISHERS, PORTABLE	1910.158							
FIRE PROTECTION EQUIPMENT					SG-13			
GASEOUS HYDROGEN SYSTEMS	1910.103							
GATE & PLUG VALVES, HIGH PRESSURE							ISDS440	
HOT WORK REPAIRS TANK BOTTOM				PSD-2207				
INDUSTRIAL PLANTS	1910.106							
INDUSTRIAL TRUCKS	1910.178							
LADDER FOR MINES			M12.1					
LIQUEFIED HYDROGEN SYSTEMS	1910.103							
METAL SCRAP PROCESSING EQUIPMENT			Z268.1					
OIL REFINERY				RP-2007				
PACKAGING EQUIPMENT			B155.1					
PIPELINE (CRUDE OIL, LPG, PRODUCTS)				PSD-2200				
POWDER COATINGS	1910.107							
POWER PRESS	1910.217							
POWER PRESS, MECHANICAL							ISDS603	
POWER TRANSMISSION APPARATUS	1910.219							
POWERED INDUSTRIAL TRUCKS	1910.178							
POWERED PLATFORMS	1910.66							
PROCESSING PLANTS	1910.106							
RESPIRATORS	1910.134							
SAFETY RELIEF DEVICES	1910.167							
(MULTIPLE REFERENCES)	1910.168							
SPRAY FINISHING	1910.107							
SPRINKLER SYSTEMS	1910.159							
STAIRS FOR MINES			M12.1					
TYPE F POWERED PLATFORMS	1910.66							
WOODWORKING MACHINERY	1910.213							
MALEIC ANHYDRIDE					SD-88			HG
MANAGEMENT POLICY							ISDS585	
EFFECTS OF FIRE ON OPERATIONS		8						
FIRE EMERGENCY		7						
MANGANESE				AQM 69-17			ISDS306	HG
TOXIC EFFECT			Z37.6					
MANHOLES	1910.23							
(MULTIPLE REFERENCES)	1926.956							
EXCAVATING	1926.956							
FLAMMABLE LIQUID		328						
GUARDING	1926.956							
OPEN FLAMES	1926.956							
TRENCHING CABLES	1926.956							

SUBJECT	USDL & USDI	NFPA	ANSI	API	MCA		NSC	MISC
	APPLICABLE STANDARDS, GUIDES AND PUBLICATIONS							
MANIFOLDS, GAS CYLINDERS	1910.252							
FUEL GASES	1910.252							
OPERATING PROCEDURES	1910.252							
PORTABLE OUTLET HEADERS	1910.252							
MANIFOLDS, OXYGEN	1926.350							
MANLIFT							ISDS401A	
CLEARANCE	1910.68							
DEFINITION	1910.68							
GENERAL REQUIREMENTS	1910.68							
HANDHOLDS	1910.68							
INSPECTION	1910.68							
INSTRUCTIONS SIGNS	1910.68							
LIGHTING	1910.68		A90.1					
MECHANICAL REQUIREMENTS	1910.68							
OPEN-PIT MINE	55.19							
OPENINGS, HOLES	1910.23							
OPERATING RULES	1910.68							
RECORDS	1910.68							
SAFETY STANDARD			A90.1					
SAND, GRAVEL, CRUSHED STONE OPERATION	56.19							
UNDERGROUND MINE	57.19							
MANTRIPS								
EQUIPMENT - GENERAL	75.1400							
MANUAL								
AIRCRAFT FIRE INVESTIGATORS		422M						
ELEVATOR INSPECTION			A17.2					
FIRE REPORTING FIELD INCIDENT		901A						
GAUGING & SAMPLING PETROLEUM TANKS							ISDS563	
MOTOR VEHICLE ACCIDENT CLASSIFICATION			D16.1					
TRAFFIC CONTROL			D6.1					
MAPS								
MINE MAPS	75.1200							
MARINE OPERATIONS	1926.605							
SEA-VALVES			Z254.1					UL 1121
THROUGH-HULL FITTINGS			Z254.1					UL 1121
WATER BUOYANT DEVICES			Z243.1					UL 1123
WORKING OVER OR NEAR WATER	1926.106							
MARINE RAILWAYS	1915.65							
(MULTIPLE REFERENCES)	1916.65							
(MULTIPLE REFERENCES)	1917.65							

SUBJECT	USDL & USDI	NFPA	ANSI	API	MCA		NSC	MISC
	APPLICABLE STANDARDS, GUIDES AND PUBLICATIONS							
MARINE TERMINALS								
MARINAS & BOATYARDS		303						
MARINE SERVICE STATIONS	1910.106							
OPERATION		307						
MARKINGS								
BULK OXYGEN SYSTEMS	1910.104							
COMPRESSED GAS CYLINDERS	1910.252							
EXPLOSIVE ACTUATED TOOLS	1910.243							
EXPLOSIVES	1910.109							
EYE & FACE PROTECTION	1910.133							
FIRE EXTINGUISHERS	1910.157							
FIRE PROTECTION EQUIPMENT	1910.144							
GASEOUS HYDROGEN SYSTEMS	1910.103							
LIQUEFIED HYDROGEN SYSTEMS	1910.103							
LOAD RATINGS, CRANES	1910.180							
LOAD RATINGS, DERRICKS	1910.181							
LOAD RATINGS, POWERED INDUSTRIAL TRUCKS	1910.178							
LOAD RATINGS, POWERED PLATFORMS	1910.66							
LPG	1910.110							
PHYSICAL HAZARDS	1910.144							
RESPIRATORS	1910.134							
STANDPIPE	1910.158							
MASONRY								
MASON'S SCAFFOLDS	1926.451							
MATERIALS HANDLING & STORAGE (SEE ALSO STORAGE)	1910.176							
(MULTIPLE REFERENCES)	1926.250							
(MULTIPLE REFERENCES)	1926.953							
(MULTIPLE REFERENCES)	1926.957							
AISLES & PASSAGEWAYS	1910.176							
ANHYDROUS AMMONIA			K61.1					CGA G-2.1
AUTOMOTIVE HOISTING EQUIPMENT							ISDS437	
BALES	1918.81							
BLASTING AGENTS			A10.7					
BREAKING DOWN	1918.83							
BRIDLES	1918.81							
BULLING	1918.84							
CASE HOOKS	1918.81							
CHAINS (ALLOY STEEL) FOR OVERHEAD LIFTING							ISDS478A	
CLEARANCE	1910.176							
CLEARANCE SIGNS	1910.176							
COMBUSTIBLES, ORDINARY	1910.177							
COMPRESSED GASES	1910.107							
CONTAINERIZED CARGO	1918.85							
CONTAINERS FOR FLAMMABLE LIQUIDS (SMALL)							ISDS532	
CONVEYORS, BELT (FOR BULK MATERIAL) - EQUIPMENT							ISDS569A	
CONVEYORS, BELT (FOR BULK MATERIAL) - OPERATION							ISDS570A	
CONVEYORS, ROLLER							ISDS528	

SUBJECT	USDL & USDI	NFPA	ANSI	API	MCA		NSC	MISC
	APPLICABLE STANDARDS, GUIDES AND PUBLICATIONS							
(CON'T)MATERIALS HANDLING & STORAGE (SEE ALSO STORAGE)								
CRANE BOOM STOPS							ISDS412	
CRANES, LOCOMOTIVE, CRAWLER & TRUCK	1910.180							
CRANES, OVERHEAD & GANTRY	1910.179							
CRANES, PENDANT-OPERATED & CONSOLE-CONTROLLED							ISDS558A	
CRAWLER, LOCOMOTIVE & TRUCK CRANES	1910.180							
DEFINITION	1910.177							
DERRICKS	1910.181							
DOCK PLATES & GANGPLANKS							ISDS318	
DRAFTS, BUILDING	1918.82							
DRAFTS, SAFELY SLUNG	1918.81							
DRAINAGE	1910.176							
EFFECTIVE DATES	1910.182							
ELECTROMAGNETS USED WITH CRANE HOISTS							ISDS359A	
EXEMPTIONS FROM STANDARD	1910.177							
EXPLOSIVES		495	A10.7					
(MULTIPLE REFERENCES)			Z271.1					
FEDERAL SUPPLY CONTRACTS	50-204.3							
FISSION MATERIAL OUTSIDE REACTOR			N16.1					
FLAMMABLE CARGO	1910.106							
FLAMMABLE LIQUID	1910.107				SG-3			
FLAMMABLE LIQUID - SHIPPING TANK		386						
FLAMMABLE LIQUID IN SMALL CONTAINERS							ISDS532	
FRAMING	1926.953							
FREIGHT CAR DOORS							ISDS374	
FRONT-END LOADER							ISDS589	
FUSES							ISDS358	
FUSES & TORPEDOES, STORAGE & HANDLING							ISDS358	
GUARDING OPENINGS	1910.176							
HAND TRUCK, POWERED							ISDS317	
HANDLING	1910.176							
HAULER-LOADER							ISDS576	
HEALTH					SG-1			
HEIGHT	1910.177							
HOIST ROPES	1926.953							
HOISTING	1918.81							
HOISTING EQUIPMENT, AUTOMATIC							ISDS437	
HOISTS, CONSTRUCTION MATERIAL							ISDS511A	
HOUSEKEEPING	1910.176							
IMPACT SENSITIVE MATERIAL					SG-7			
INDOOR STORAGE	1910.177							
LIQUIDS, DIP TANK	1910.108							
LIQUIDS, HANDLING FOR DIP TANKS	1910.108							
LNG		59A	Z225.1					
LOAD ATTACHING	1926.953							
LPG	1910.110	58	Z106.1					
LPG FOR INDUSTRIAL TRUCKS							ISDS479A	
LUMBER	1918.81							
MAGNESIUM		48						
MARINE OPERATIONS	1926.605							

SUBJECT	USDL & USDI	NFPA	ANSI	API	MCA		NSC	MISC
			APPLICABLE STANDARDS, GUIDES AND PUBLICATIONS					
(CON'T)MATERIALS HANDLING & STORAGE (SEE ALSO STORAGE)								
MECHANICAL EQUIPMENT	1910.176							
MOTOR TRUCK FOR MINES, QUARRIES & CONSTRUCTION							ISDS330A	
NUCLEAR POWER PLANT MATERIALS			N45.2.2					
OIL FIELD PIPE, HANDLING LARGE DIAMETER							ISDS463	
OIL FILLED EQUIPMENT	1926.953							
OPEN-PIT MINE	55.16							
OVERHEAD GANTRY CRANES	1910.179							
PACKAGING EQUIPMENT - CONSTRUCTION, CARE, USE			B155.1					
PAPER BAG & SACK MANUFACTURE			Z259.1					
PERSONAL PROTECTION	1910.176							
PETROLEUM PRODUCTS FIRE PROTECTION				RP-2001				
PIPES	1918.81							
PLASTIC FOAM					SG-5			
POLE HAULING	1926.953							
POWER PRESS, HANDLING FINISHED PIECES							ISDS534	
POWERED INDUSTRIAL TRUCKS	1910.178							
RADIOACTIVE MATERIALS		801						
RAILROAD CAR	1918.84							
RAILROAD CAR BLOCKS	1910.176							
REEFERS	1918.83							
SAFETY BELTS	1926.250							
SAND, GRAVEL, CRUSHED STONE OPERATION	56.16							
SCRAP BAILERS							ISDS611	
SECURING	1910.176							
(MULTIPLE REFERENCES)	1918.83							
SHAFTS & CORES, HANDLING							ISDS361	
SHEET METAL, HANDLING & STORAGE							ISDS434	
SHOCK SENSITIVE MATERIAL					SG-7			
SKIDS							ISDS260A	
STANDARDS SOURCES	1910.183							
STEEL PLATE, HANDLING FOR FABRICATION							ISDS565	
STORAGE	1910.176							
(MULTIPLE REFERENCES)	1926.953							
STORES, FIRE PREVENTION							ISDS549	
SULFUR, HANDLING & STORAGE OF SOLID							ISDS612	
SULFUR, HANDLING LIQUID							ISDS592	
TAG LINES	1918.81							
(MULTIPLE REFERENCES)	1926.953							
THAWING FROZEN LADINGS							ISDS233	
TITANIUM		481						
TORPEDO							ISDS358	
TRIMMING	1918.83							
TYPE I	1910.177							
TYPE II	1910.177							
TYPE III	1910.177							
UNDERGROUND MINE	56.16							
UNLOADING	1926.953							
WATER PIPES & FITTINGS							ISDS623	
WINCHES, TRUCK MOUNTED POWER							ISDS441	

SUBJECT	USDL & USDI	NFPA	ANSI	API	MCA		NSC	MISC
		APPLICABLE STANDARDS, GUIDES AND PUBLICATIONS						
(CON'T)MATERIALS HANDLING & STORAGE (SEE ALSO STORAGE)								
WIRE ROPE SLINGS, RECOMMENDED LOADS							ISDS380	
MATS	1910.137							
MEANS OF EGRESS	1910.35							
ACCESS TO EXITS	1910.37							
AISLES & PASSAGEWAYS	1910.22							
ALARM & DETECTION SYSTEMS, FIRE	1910.37							
APPLICATION	1910.36							
AUTOMATIC SPRINKLER SYSTEMS	1910.37							
CAPACITY	1910.37							
CHANGE IN ELEVATION	1910.37							
CONSTRUCTION WORKERS	1910.36							
DEFINITION	1910.35							
DISCHARGE FROM EXITS	1910.37							
DOORS	1910.252							
EGRESS CAPACITY & OCCUPANT LOAD	1910.37							
EMPLOYEE PROTECTION	1910.36							
ENCLOSED SPACE					SG-10			
EXIT COMPONENTS	1910.37							
EXIT MARKINGS	1910.37							
EXTERIOR WAYS TO EXIT ACCESS	1910.37							
FURNISHINGS & DECORATIONS	1910.37							
HEADROOM	1910.37							
MAINTENANCE & WORKMANSHIP	1910.37							
PASSAGEWAYS	1910.22							
(MULTIPLE REFERENCES)	1910.36							
PERMISSIBLE EXIT COMPONENTS	1910.37							
PETROLEUM STORAGE TANK				RP-2015				
PROTECTIVE ENCLOSURE OF EXITS	1910.37							
RAMP ENCLOSURES	1910.23							
RAMP EXITS	1910.37							
REPAIR OPERATIONS	1910.36							
STANDARDS SOURCES	1910.39							
TANKS					SG-10			
MEASUREMENT								
AUDIOMETER SPECIFICATIONS			S3.6					
ELECTRICAL			C39.5					
SOUND			S1.2					
SOUND - PNEUMATIC EQUIPMENT			S5.1					
SOUND - PRESSURE LEVEL			S1.13					
SOUND LEVEL METERS			S1.4					
MECHANICAL EQUIPMENT	1926.952							
ABRASIVE BLASTING PAINT REMOVER	1915.23							
(MULTIPLE REFERENCES)	1916.23							
AERIAL LIFTS	1926.952							
CLEARANCE	1910.176							

SUBJECT	USDL & USDI	NFPA	ANSI	API	MCA		NSC	MISC
	APPLICABLE STANDARDS, GUIDES AND PUBLICATIONS							
(CON'T)MECHANICAL EQUIPMENT								
CRANES	1926.952							
DERRICKS	1926.952							
ENERGIZED SUBSTATIONS	1926.957							
FLAME PAINT REMOVER	1915.23							
(MULTIPLE REFERENCES)	1916.23							
HANDLING	1910.176							
INSPECTION	1926.952							
MATERIALS HANDLING	1910.177							
MATERIALS HANDLING & STORAGE	1910.176							
PERSONAL PROTECTION - PAINT REMOVER	1915.23							
(MULTIPLE REFERENCES)	1916.23							
POWERED INDUSTRIAL TRUCKS	1910.177							
MECHANICAL POWER TRANSMISSION APPARATUS	1910.219							
MEDICAL (SEE ALSO INDUSTRIAL HYG &/OR FIRST AID)								
ATTENTION	1926.23							
BENZIDINE - EXAMINATIONS & RECORDS	1910.93							
EXAMINATIONS, ASBESTOS	1910.93							
FEDERAL SUPPLY CONTRACTS	50-204.5							
FUMIGATION GAS SYSTEM		57						
INHALATION ANESTHETICS GAS SYSTEM		56A						
INHALATION THERAPY GAS SYSTEM		56B						
NONFLAMMABLE GAS SYSTEM		56F						
RECORDS - ASBESTOS	1910.93							
SERVICES	1926.50							
TREATMENT	1910.151							
MERCANTILE OCCUPANCIES	1910.106							
MERCURY	1910.252			AQM 69-16				HG
EXPOSURE LIMIT	1910.95							
FULMINATE OF MERCURY							ISDS309A	
INDUSTRIAL SAFETY DATA SHEET							ISDS203	
TOXIC EFFECT			Z37.8					
MERCURY, ORGANO (ALKYL)								
TOXIC EFFECT			Z37.30					
MESITYL OXIDE								HG
METAL & NONMETALLIC OPEN PIT MINES	55.1							
AERIAL TRAMWAYS	55.10							
AIR QUALITY	55.5							
COMPRESSED AIR & BOILERS	55.13							
DEFINITION	55.2							
DRILLING	55.7							
ELECTRICITY	55.12							
EXPLOSIVES	55.6							

SUBJECT	USDL & USDI	NFPA	ANSI	API	MCA		NSC	MISC
	APPLICABLE STANDARDS, GUIDES AND PUBLICATIONS							
(CON'T)METAL & NONMETALLIC OPEN PIT MINES								
FIRE PREVENTION & CONTROL	55.4							
GROUND CONTROL	55.3							
ILLUMINATION	55.17							
LOADING, HAULING, DUMPING	55.9							
MAN HOISTING	55.19							
MATERIALS HANDLING & STORAGE	55.16							
PERSONAL PROTECTION	55.15							
ROTARY JET PIERCING	55.8							
SAFETY PROGRAMS	55.18							
TRAVELWAYS	55.11							
USE OF EQUIPMENT	55.14							
METAL & NONMETALLIC UNDERGROUND MINES	57.1							
AERIAL TRAMWAYS	57.10							
AIR QUALITY	57.5							
COMPRESSED AIR & BOILERS	57.13							
DEFINITION	57.2							
DRILLING	57.7							
ELECTRICITY	57.12							
EXPLOSIVES	57.6							
FIRE PREVENTION & CONTROL	57.4							
GROUND CONTROL	57.3							
ILLUMINATION	57.17							
LOADING, HAULING, DUMPING	57.9							
MAN HOISTING	57.19							
MATERIALS HANDLING & STORAGE	57.16							
PERSONAL PROTECTION	57.15							
RADIATION	57.5							
ROTARY JET PIERCING	57.8							
SAFETY PROGRAMS	57.18							
TRAVELWAYS & ESCAPEWAYS	57.11							
USE OF EQUIPMENT	57.14							
VENTILATION	57.5							
METAL HYDRIDES							ISDS462	
METAL SCRAP								
PROCESSING EQUIPMENT - DESIGN, USE, MAINTENANCE			Z268.1					
METAL TOWER CONSTRUCTION	1926.955							
CRANES	1926.955							
FOOTINGS	1926.955							
FOOTINGS, AUGER-TYPE	1926.955							
FOOTINGS, EXCAVATIONS	1926.955							
FOOTINGS, PAD-TYPE	1926.955							
FOOTINGS, PILE-TYPE	1926.955							
GUY LINES	1926.955							
SUPPORTS	1926.955							

SUBJECT	APPLICABLE STANDARDS, GUIDES AND PUBLICATIONS							
	USDL & USDI	NFPA	ANSI	API	MCA		NSC	MISC
METAL WORKING								
DRILL PRESS							ISDS335	
MILLING MACHINE							ISDS364	
METALIZING							ISDS357	
METHANAL (SEE FORMALDEHYDE)								
METHANE								
TEST	75.150							
(MULTIPLE REFERENCES)	75.151							
METHANOL					CC-69			
(MULTIPLE REFERENCES)					SD-22			
(MULTIPLE REFERENCES)					TEG		ISDS407A	HG
ELECTRICAL METAL-WORKING MACHINE TOOLS		79						
TOXIC EFFECT			Z37.14					
2-METHOXYETHANOL								HG
METHYL ACETATE								HG
METHYL ACRYLATE					CC-74			
(MULTIPLE REFERENCES)					CIC-58			
(MULTIPLE REFERENCES)					SD-79			
(MULTIPLE REFERENCES)					TEG			
METHYL ALCOHOL (SEE METHANOL)								
METHYLAMINES					SD-57			
METHYLAMINES (ANHYDROUS)					CC-26			
(MULTIPLE REFERENCES)					TEG			
METHYLAMINES (AQUEOUS)					CC-84			
(MULTIPLE REFERENCES)					TEG			
METHYLBENZENE (SEE TOLUENE)								
METHYL BROMIDE (SEE BROMOMETHANE)								
METHYL CELLOSOLVE (SEE 2-METHOXYETHANOL)								
METHYL CHLORIDE (SEE CHLOROMETHANE)								
METHYL CHLOROFORM (SEE TRICHLOROETHANE)								

SUBJECT	USDL & USDI	NFPA	ANSI	API	MCA		NSC	MISC
	APPLICABLE STANDARDS, GUIDES AND PUBLICATIONS							
METHYL CHLOROMETHYL ETHER	1910.93							
METHYL DICHLORIDE (SEE DICHLOROMETHANE)								
4,4'-METHYLENE BIS(2-CHLOROANILINE)	1910.93							
METHYLENE CHLORIDE (SEE DICHLOROMETHANE)								
METHYL ETHYL KETONE (SEE 2-BUTANONE)								
METHYL ISOBUTYL KETONE					CC-57			
(MULTIPLE REFERENCES)					TEG			HG
METHYL METHACRYLATE					CC-43			
(MULTIPLE REFERENCES)					CIC-61			
(MULTIPLE REFERENCES)					TEG			
MIDRAILS								
SCAFFOLDING	1926.451							
SPECIFICATIONS	1926.500							
MILLING MACHINE								
METALWORKING							ISDS364	
MILLS & CALENDERS								
ALARMS	1910.216							
FLOUR & FEED - DUST EXPLOSION		61C	Z12.3					
PLASTICS INDUSTRY	1910.216							
PROTECTED BY LOCATION	1910.216							
PULP, PAPER & PAPERBOARD			P1.1					
RUBBER & PLASTICS	1910.216		B28.1					
SAFETY CONTROLS	1910.216							
STOPPING LIMITS	1910.216							
TRIP & ENERGIZING SWITCHES	1910.216							
MINERAL SEAL OIL				TR				
MINING								
COAL DUST - RESPIRATORY PROTECTION			Z88.4					
LADDERS - CONSTRUCTION & MAINTENANCE			M12.1					
RADIATION STANDARDS	1910.96							
STAIRS - CONSTRUCTION & MAINTENANCE			M12.1					
URANIUM - RADIATION PROTECTION			N13.8					
(MULTIPLE REFERENCES)			N7.1					
WIRE ROPE - SPECIFICATIONS & USE			M11.1					

SUBJECT	USDL & USDI	NFPA	ANSI	API	MCA		NSC	MISC
	APPLICABLE STANDARDS, GUIDES AND PUBLICATIONS							
MINORS, EMPLOYMENT	1910.217							
(MULTIPLE REFERENCES)	1918.97							
MISTS	1910.93							
(MULTIPLE REFERENCES)	1926.55							
FEDERAL SUPPLY CONTRACTS	50-204.50							
IN INDUSTRY							ISDS531	
MIXED ACID					CC-65			
(MULTIPLE REFERENCES)					SD-65			
(MULTIPLE REFERENCES)					TEG			
MIXERS & MULLERS	1910.263							
PAN-TYPE							ISDS526A	
MOBILE HOMES								
DESIGN & CONSTRUCTION			A119.1					
HEATING APPLIANCES USING LIQUID FUEL			A147.1					UL 307A
OCCUPANCY STANDARDS & PROCESS HAZARDS		501B						
PARKS FOR MOBILE HOMES		501A						
ROOF JACK			A160.1					UL 311
MOBILE LADDERS, STANDS & SCAFFOLDS, MANUALLY PROP	1910.29							
BASE WIDTH	1910.29							
GENERAL REQUIREMENTS	1910.29							
HANDRAILS	1910.29							
LOADING	1910.29							
MOBILE LADDER STANDS	1910.29							
MOBILE TUBE & COUPLER SCAFFOLDS	1910.29							
MOBILE TUBULAR WELDED FRAME SCAFFOLDS	1910.29							
MOBILE TUBULAR WELDED SECTIONAL FOLDING SCAFFOLD	1910.29							
MOBILE WORK PLATFORMS	1910.29							
STEPS	1910.29							
WHEELS OR CASTERS	1910.29							
WORK LEVELS	1910.29							
WORKING LOADS	1910.29							
MOBILE SCAFFOLDS, TUBULAR WELDED	1910.29							
BRACING	1910.29							
END FRAMES	1910.29							
ERECTION	1910.29							
LOCKING	1910.29							
MANUALLY PROPELLED	1926.451							
SECTIONAL FOLDING LADDER	1910.29							
SECTIONAL FOLDING LADDER STAIRWAY	1910.29							
SPACING	1910.29							
STAIRWAYS	1910.29							

APPLICABLE STANDARDS, GUIDES AND PUBLICATIONS

SUBJECT	USDL & USDI	NFPA	ANSI	API	MCA		NSC	MISC
MOBILE TUBE & COUPLER	1910.29							
DESIGN	1910.29							
ERECTION	1910.29							
MATERIALS	1910.29							
MOBILE WORK PLATFORMS	1910.29							
BASE WIDTH	1910.29							
BRACING	1910.29							
MODEL ROCKETRY								
FIRE CODE		41L						
MODEL STATE LAW								
INSPECTION - ELECTRICAL SYSTEMS		70L						
MOLDING								
COMPRESSION, PLASTICS							ISDS632	
INJECTION							ISDS454A	
MACHINERY	1910.213							
MOLYBDENUM								HG
MONITORING								
ASBESTOS	1910.93							
IONIZING RADIATION	1910.96							
RECORDS	1910.93							
(MULTIPLE REFERENCES)	1910.96							
MONOBROMOMETHANE (SEE BROMOMETHANE)								
MONOCHLORODIFLUOROMETHANE					CIC-62			
MONOETHANOLAMINE					CIC-63			
MONOISOPROPANOLAMINE					CIC-64			
MONOMETHYL HYDRAZINE					CC-11			
(MULTIPLE REFERENCES)					TEG			
MONORAIL SYSTEMS								
SAFETY STANDARD			B30.11					
MORPHOLINE					CIC-65			
MOTION PICTURE								
SAFETY							ISDS556A	

SUBJECT	USDL & USDI	NFPA	ANSI	API	MCA		NSC	MISC
	APPLICABLE STANDARDS, GUIDES AND PUBLICATIONS							
MOTOR CRAFT								
FIRE SIGNAL		311						
GAS HAZARD CONTROL		306						
PROTECTION DURING CONSTRUCTION		312						
SAFETY STANDARD		302	Z120.1					
MOTORCYCLE								
INSPECTION			D7.5					
MOTOR FUEL ANTIKNOCK COMPOUND					CC-30			
(MULTIPLE REFERENCES)					CIC-66			
(MULTIPLE REFERENCES)					TEG			
MOTOR FUEL LPG	1910.110							
MOTOR VEHICLES	1926.600							
(MULTIPLE REFERENCES)	1926.601							
ACCIDENT - CLASSIFICATION MANUAL			D16.1					
ACCIDENT - RECORDING & MEASURING			D15.1					
(MULTIPLE REFERENCES)			D15.2					
AIRCRAFT RESCUE & FIRE FIGHTING VEHICLE		414	B128.1					
ANHYDROUS AMMONIA	1910.111							
ELECTRICAL STANDARDS - RECREATIONAL VEHICLE PARK		501D						
GANTRY TRUCK							ISDS257	
GLAZING MATERIAL			Z26.1					
INDUSTRIAL TRUCKS - POWERED			B56.1					ISO R1074
INDUSTRIAL TRUCKS - POWERED, USE, MAINT, OPERATN		505	B56.2					
INSPECTION			D7.1					
INSPECTION STATION			D7.2					
(MULTIPLE REFERENCES)			D7.3					
MOBILE HOME - HEATING APPL USING LIQUID FUEL			A147.1					UL 307A
MOBILE HOME - OCCUPANCY STANDARD & PROC HAZARD		501B						
RECREATIONAL - OCCUPANCY STANDARDS & PROC HAZARD		501C						
RECREATIONAL - SAFETY STANDARD			A119.2					
ROAD LIGHTING			D12.1					IES RP8
TRAFFIC CONTROL - ADJUSTABLE FACE SIGNAL			D10.1					ITE TR1
TRAFFIC CONTROL - PRE-TIMED, FIXED SIGNAL			D11.1					ITE TR2
TRAFFIC CONTROL - TRAFFIC ACTUATED			D13.1					ITE TR3
TRAFFIC CONTROL MANUAL			D6.1					
MOULDERS	1910.263							
MOUNTING								
TIRES & RIMS, HEAVY DUTY							ISDS411	
MOVING WALK								
SAFETY CODE			A17.1					

SUBJECT	USDL & USDI	NFPA	ANSI	API	MCA		NSC	MISC
MUD FLAPS	1926.601							
MULLERS & MIXERS PAN-TYPE							ISDS526A	
MURIATIC ACID (SEE HYDROCHLORIC ACID)								

SUBJECT	USDL & USDI	NFPA	ANSI	API	MCA		NSC	MISC
	APPLICABLE STANDARDS, GUIDES AND PUBLICATIONS							
NAILS, PROTRUDING	1918.91							
NAPHTHA								
PETROLEUM				TR				
PETROLEUM (AROMATIC)				TR				
NAPHTHALENE				TR	SD-58		ISDS370A	HG
NAPHTHENIC ACID				TR				
ALPHA-NAPHTHYLAMINE	1910.93							
BETA-NAPHTHYLAMINE	1910.93				SD-32			
NATIONAL CONSENSUS STANDARDS	1910.2							
NATIONAL ELECTRICAL CODE (MULTIPLE REFERENCES)	1910.308 1910.309							
NETS, SAFETY	1926.105							
CONSTRUCTION							ISDS608	
GANGWAYS	1918.21							
SAFETY	1918.32							
STEEL ERECTION	1926.750							
NEUTRON-MULTIPLICATION MEASUREMENTS IN-SITU			N16.3					
NICKEL				AQM 69-14				HG
NICKEL CARBONYL								HG
NITRIC ACID					CC-47			
(MULTIPLE REFERENCES)					SD-5			
(MULTIPLE REFERENCES)					TEG			HG
NITRIC ACID (RED, FUMING)					CC-3			
(MULTIPLE REFERENCES)					TEG			
PARA-NITROANILINE					SD-94			
NITROBENZENE					CC-79			
(MULTIPLE REFERENCES)					SD-21			
(MULTIPLE REFERENCES)					TEG			HG
4-NITROBIPHENYL	1910.93							

SUBJECT	USDL & USDI	NFPA	ANSI	API	MCA		NSC	MISC
	APPLICABLE STANDARDS, GUIDES AND PUBLICATIONS							
NITROCELLULOSE					SD-96			
NITROETHANE								HG
NITROGEN (LIQUID) (MULTIPLE REFERENCES)					CC-6 TEG			
NITROGEN DIOXIDE								HG
TOXIC EFFECT			Z37.13					
NITROGEN OXIDES							ISDS206	
NITROGEN TETROXIDE (MULTIPLE REFERENCES)					CC-1 TEG			
NITROGLYCERINE								HG
NITROMETHANE								HG
NITROPROPANE								HG
N-NITROSODIMETHYLAMINE	1910.93							
NITROUS OXIDE	1910.105							
FEDERAL SUPPLY CONTRACTS	50-204.69							
NOISE								
AUDIOMETER ROOM			S3.1					
EXPOSURE (MULTIPLE REFERENCES)	1910.95 1926.52							
FEDERAL SUPPLY CONTRACTS	50-204.10							
GUIDELINES				TB-7301				
HEARING LOSS TO NOISE EXPOSURE			Z24-X-2					
JET ENGINE							ISDS580	
LOUDNESS COMPUTATION			S3.4					ISO R357
PROTECTIVE EQUIP	1910.95							
STANDARDS (MULTIPLE REFERENCES)	70.500 71.300							
NONDESTRUCTIVE EXAMINATIONS								
ACCREDITED PERSONS	1919.78							
DEFINITION	1919.2							
SHORE-BASED EQUIPMENT	1919.50							
NONIONIZING RADIATION (MULTIPLE REFERENCES)	1910.97 1926.54							
ELECTROMAGNETIC RADIATION	1910.97							

SUBJECT	USDL & USDI	NFPA	ANSI	API	MCA		NSC	MISC
	APPLICABLE STANDARDS, GUIDES AND PUBLICATIONS							
NONWATER CARRIAGE DISPOSAL SYSTEMS	1910.143							
ACCEPTABLE INDUSTRIAL DISPOSAL SYSTEMS	1910.143							
CHEMICAL TOILET SPECIFICATIONS	1910.143							
COMBINATION TOILETS	1910.143							
PORTABLE TOILET CONSTRUCTION	1910.143							
PRIVY SPECIFICATIONS	1910.143							
RECIRCULATING TOILET SPECIFICATIONS	1910.143							
SEEPAGE PIT CONSTRUCTION	1910.143							
NOTIFYING SHIPS OFFICERS	1918.75							
NOXIOUS GASES, STORAGE AREAS	1910.178							
NOZZLES								
ABRASIVE BLASTING	1910.244							
AIR CLEANING	1910.94							
NUCLEAR ENERGY								
INFORMATION SOURCE							ISDS446	
NUCLEAR FUEL PLANT								
RADIATION PROTECTION			N7.2					
NUCLEAR POWER								
ELECTRICAL EQUIP INSTALLATION, INSPECTION, TESTG			N45.2.4					IEEE 336
GENERATING STATION RADIATION PROTECTION			N42.7					IEEE 279
INSTRUMENT INSTALLATION, INSPECTION, TESTING			N45.2.4					IEEE 336
MTRLS HANDLING/STORAGE/PACKAGE/SHIP/RECEIVE			N45.2.2					
PIPING			B31.7					
NUCLEAR REACTOR								
HANDLING FISSIONABLE MATERIAL OUTSIDE REACTOR			N16.1					
PROCESS HAZARDS		802						

APPLICABLE STANDARDS, GUIDES AND PUBLICATIONS

SUBJECT	USDL & USDI	NFPA	ANSI	API	MCA		NSC	MISC
OBSTRUCTIONS, EXIT ACCESS	1910.37							
OCCUPATIONAL HEALTH & ENVIRONMENTAL CONTROL	1926.50							
ADDITIONAL DELAY IN EFFECTIVE DATE	1910.98							
AIR CONTAMINANTS	1910.93							
FOOT PROTECTION	1910.136							
HEAD PROTECTION	1910.135							
IONIZING RADIATION	1910.96							
MANAGEMENT POLICY							ISDS585	
NONIONIZING RADIATION	1910.97							
OCCUPATIONAL NOISE EXPOSURE	1910.95							
(MULTIPLE REFERENCES)	1926.52							
STANDARDS ORGANIZATIONS	1910.100							
STANDARDS SOURCES	1910.99							
VENTILATION	1910.94							
ODORIZING GASES	1910.110							
OFFICE EQUIPMENT								
SAFETY STANDARD			X4.12					UL 114
OFFICE OCCUPANCIES	1910.106							
OFFSHORE PRODUCTION								
DIVING SAFETY				TB-7104				
OIL BURNING EQUIPMENT								
CENTRAL FURNACE			Z96.1					UL 727
SAFETY CONTROLS			Z226.1					UL 372
SAFETY STANDARD		31	Z95.1					
OIL FIELD								
HANDLING PIPE							ISDS463	
OIL FILLED EQUIPMENT	1926.953							
OIL MIST								
SAMPLING PROCEDURE EVALUATION				TB-7201				
OIL REFINERY								
FIRE PROTECTION				RP-2001				
MAINTENANCE				RP-2007				
PIPING			B31.3					
WORKERS MORTALITY STUDY - TABERSHAW COOPER ASSOC				MRR-7402				
OIL WELL								
FRACTURING WITH FLAMMABLE LIQUID				PSD-2204				

APPLICABLE STANDARDS, GUIDES AND PUBLICATIONS

SUBJECT	USDL & USDI	NFPA	ANSI	API	MCA		NSC	MISC
OILING FACILITIES, POWER TRANSMISSION	1910.219							
OLEUM (SEE SULFURIC ACID)								
OPEN-PIT MINE								
SAFETY	55.1							
OPEN-SIDED FLOORS, PROTECTION	1910.23							
OPEN-SURFACE TANKS								
VENTILATION	1910.94		Z9.1					
OPENINGS	1910.23							
FLOORS & WALLS	1926.500							
TANKS, ABOVE GROUND	1910.106							
TANKS, INSIDE	1910.106							
WALL PROTECTION	1910.23							
OPERATION								
DERRICKS	1910.181							
DIP TANKS		34						
INDUSTRIAL TRUCKS	1910.178							
INDUSTRIAL TRUCKS, TRAINING	1910.178							
POWER PRESS	1910.217							
SPRAY FINISHING	1910.107							
WELDING & CUTTING	1910.251							
OPERATOR								
ABRASIVE WHEELS	1915.74							
(MULTIPLE REFERENCES)	1916.74							
COMPETENT PERSONS	1915.67							
(MULTIPLE REFERENCES)	1916.67							
INCAPACITATING AILMENTS	1915.67							
(MULTIPLE REFERENCES)	1916.67							
(MULTIPLE REFERENCES)	1917.67							
MINORS	1915.67							
(MULTIPLE REFERENCES)	1916.67							
(MULTIPLE REFERENCES)	1917.57							
SIGNS	1915.67							
(MULTIPLE REFERENCES)	1916.67							
(MULTIPLE REFERENCES)	1917.67							
ORGANIC PEROXIDES	1910.107	33						
ORGANO (ALKYL) MERCURY								
TOXIC EFFECT			Z37.30					

APPLICABLE STANDARDS, GUIDES AND PUBLICATIONS

SUBJECT	USDL & USDI	NFPA	ANSI	API	MCA		NSC	MISC
OSMIUM & ITS COMPOUNDS								HG
OUTLETS, FIRE HOSE	1910.158							
OUTLET HEADERS	1910.252							
OVENS	1910.263							
OVERBOARD DISCHARGES	1918.94							
OVERCURRENT PROTECTION	1926.402							
OVERHEAD GANTRY CRANES	1910.179							
BRAKES	1910.179							
CABS	1910.179							
DEFINITION	1910.179							
ELECTRICAL EQUIPMENT	1910.179							
FOOTWALKS & LADDERS	1910.179							
GENERAL REQUIREMENTS	1910.179							
HANDLING LOAD	1910.179							
HOISTING EQUIPMENT	1910.179							
INSPECTION	1910.179							
MAINTENANCE	1910.179							
OTHER REQUIREMENTS	1910.179							
ROPE INSPECTION	1910.179							
STOPS, BUMPERS, RAIL SWEEPS & GUARDS	1910.179							
TESTING	1910.179							
OVERHEAD LINES	1926.955							
CLEARANCE	1926.955							
CLIMBING ELEVATED STRUCTURES	1926.955							
CONDUCTOR GRIPS	1926.955							
DEENERGIZED CONDUCTORS	1926.955							
ERECTION CRANES	1926.955							
INSPECTION	1926.955							
LIFTING EQUIPMENT	1926.955							
LOADLINES	1926.955							
POLE HOLES	1926.955							
REEL HANDLING EQUIPMENT	1926.955							
SPLICING	1926.965							
STRINGING	1926.955							
TAG LINES	1926.955							
TOWER ASSEMBLY	1926.955							
TRAFFIC CONTROL	1926.955							
TRANSMISSION CLIPPING	1926.955							
WIND, BAD WEATHER	1926.955							

SUBJECT	USDL & USDI	NFPA	ANSI	API	MCA		NSC	MISC
OVERHEAD PROTECTION	1926.451							
(MULTIPLE REFERENCES)	1926.1003							
OVERHEAD WIRES								
CRANES	1910.180							
DERRICKS	1910.181							
OVERSPRAY COLLECTORS	1910.107							
OXALIC ACID							ISDS406	
OXIDANTS								
AIR QUALITY CRITERIA & HEALTH				AQM 69-5				
OXIDIZING MATERIAL								
LIQUID		43A						
SOLID		43A						
OXYGEN	1910.104				CC-13			
CONCENTRATION, DIP TANK	1910.94							
(MULTIPLE REFERENCES)	1910.108							
DEFICIENCY TEST	75.150							
FEDERAL SUPPLY CONTRACTS	50-204.67							
GASEOUS							ISDS472	
LIQUID					TEG			
OXYGEN DIFLUORIDE								HG
OXYGEN MANIFOLDS	1926.350							
OXYGEN SYSTEMS								
BULK - CLEANING	1910.104							
BULK - CONTAINERS	1910.104							
BULK - DIKES	1910.104							
BULK - DISTANCE TO OTHER EXPOSURE	1910.104							
BULK - ELECTRIC WIRING	1910.104							
BULK - FIRE CODE		50						
BULK - LOCATION	1910.104							
BULK - MAINTENANCE	1910.104							
BULK - MARKINGS	1910.104							
BULK - SAFETY RELIEF DEVICES	1910.104							
BULK - TESTING	1910.104							
CUTTING		51						
WELDING		51						
OZONE								HG

SUBJECT	USDL & USDI	NFPA	ANSI	API	MCA		NSC	MISC
	APPLICABLE STANDARDS, GUIDES AND PUBLICATIONS							
PACKAGES	1910.177							
PACKAGING EQUIPMENT								
CONSTRUCTION, CARE, USE			B155.1					
PAINT REMOVAL								
CHEMICAL	1915.22							
(MULTIPLE REFERENCES)	1916.22							
MECHANICAL	1915.23							
(MULTIPLE REFERENCES)	1916.23							
PAINTING, COATINGS								
AIRLESS SPRAY PAINTING							ISDS548	
ELECTROSTATIC PAINT SPRAYING							ISDS468	
PAINTS IN FLAMMABLE SOLVENTS	1915.24							
(MULTIPLE REFERENCES)	1916.24							
PAINTS WITH TOXIC SOLVENTS	1915.24							
(MULTIPLE REFERENCES)	1916.24							
PAINTS, FIRE RETARDANT	1910.37							
PAINTS, TOXICITY SPECIFICATIONS			Z66.1					
PAPER BAG, SACK								
MANUFACTURE - SAFETY/HEALTH REQUIREMENTS			Z259.1					
PAPER MILL	1910.261							
SAFETY STANDARD			P1.1					
PAPERBOARD MILL	1910.261							
SAFETY STANDARD			P1.1					
PARATHION								HG
PARTICLE ACCELERATOR								
RADIOLOGICAL SAFETY			N43.1					NBS H107
PARTICULATES								
AIR QUALITY CRITERIA & HEALTH				AQM 69-2				
AMBIENT AIR QUALITY STANDARDS				AQM 69-1				
PASSAGEWAYS, WALKWAYS, RUNWAYS	1910.22							
(MULTIPLE REFERENCES)	1910.36							
(MULTIPLE REFERENCES)	1926.500							
CRANES & DERRICKS	1926.550							
ELECTRICAL EQUIPMENT	1926.400							
PAVING								
PORTLAND CEMENT CONCRETE							ISDS541	

SUBJECT	USDL & USDI	NFPA	ANSI	API	MCA		NSC	MISC
	APPLICABLE STANDARDS, GUIDES AND PUBLICATIONS							
PENTABORANE (MULTIPLE REFERENCES)					CC-5 TEG			HG
PENTACHLOROPHENOL								HG
PENTAERYTHRITOL								HG
PENTANE								HG
PERACETIC ACID (MULTIPLE REFERENCES)					CC-86 TEG			
PERCHLORIC ACID							ISDS311A	
PERCHLORIC ACID SOLUTION					SD-11			
PERCHLOROETHYLENE					SD-24			
PERCHLORYL FLUORIDE (MULTIPLE REFERENCES)					CC-4 TEG			
PEROXIDES, ORGANIC	1910.107							
PERSONAL CLIMBING EQUIP (SEE CLIMBING EQUIPMENT)								
PERSONAL PROTECTIVE EQUIP (SEE PROTECTIVE EQUIPMENT)								
PEST CONTROL	1910.141							
PETROLEUM ASPHALT								
HEALTH ASPECTS				TB-7103				
PETROLEUM GASES	1910.110							
CONTAINER UTILIZATION	1910.110							
CYLINDER SYSTEMS	1910.110							
GAS SERVICE STATIONS	1910.110							
MOTOR FUELS	1910.110							
ODORIZING	1910.110							
RESALE STORAGE CONTAINERS	1910.110							
PETROLEUM NAPHTHA				TR				HG
PETROLEUM PRODUCTS								
FLAME ARRESTOR - STORAGE TANK VENT			Z222.1					UL 525
PIPELINE REPAIR GUIDE				PSD-2200				

APPLICABLE STANDARDS, GUIDES AND PUBLICATIONS

SUBJECT	USDL & USDI	NFPA	ANSI	API	MCA		NSC	MISC
PETROLEUM TANKS								
FLAME ARRESTOR - PETROLEUM OIL/GASOLINE			Z222.1					UL 525
GAUGING							ISDS563	
SAMPLING							ISDS563	
PHENOL					CC-48			
(MULTIPLE REFERENCES)					CIC-68			
(MULTIPLE REFERENCES)					SD-4			
(MULTIPLE REFERENCES)					TEG		ISDS405A	HG
PHOSGENE					SD-95			HG
PHOSPHINE								HG
PHOSPHORIC ACID					CIC-69			
(MULTIPLE REFERENCES)					SD-70			HG
TANK CAR UNLOADING					TC-2			
PHOSPHORIC ANHYDRIDE					SD-28			HG
PHOSPHORUS, ELEMENTAL				AQM 69-19	CC-37			
(MULTIPLE REFERENCES)					CIC-70			
(MULTIPLE REFERENCES)					SD-16			
(MULTIPLE REFERENCES)					TEG			
PHOSPHORUS, WHITE							ISDS282A	
PHOSPHORUS OXYCHLORIDE					CC-39			
(MULTIPLE REFERENCES)					SD-26			
(MULTIPLE REFERENCES)					TEG			
PHOSPHORUS PENTASULFIDE					CC-41			
(MULTIPLE REFERENCES)					SD-71			
(MULTIPLE REFERENCES)					TEG			
PHOSPHORUS TRICHLORIDE					CC-40			
(MULTIPLE REFERENCES)					SD-27			
(MULTIPLE REFERENCES)					TEG			
PHOTOCHEMICAL SMOG				AQM 69-6				
PHOTOGRAPHY								
SAFETY PROFESSIONAL							ISDS619	
PHTHALIC ANHYDRIDE					SD-61			HG

SUBJECT	USDL & USDI	NFPA	ANSI	API	MCA		NSC	MISC
			APPLICABLE STANDARDS, GUIDES AND PUBLICATIONS					
PHYSICAL HAZARDS, COLOR CODES	1910.144							
PICKLING								
STEEL BILLETS, BLOOMS, RODS & BARS							ISDS453	
PICRIC ACID							ISDS351A	
PIERS & WHARVES								
OCCUPANCY STANDARDS & PROCESS HAZARDS		87						
PIER HOLES, TRENCHING	1926.652							
PILE DRIVING	1926.603							
PILING, MATERIALS HANDLING	1910.176							
INDOOR STORAGE	1910.177							
PIPE								
CONNECTORS FOR FLAMMABLE LIQUIDS, LPG			B148.1					UL 567
FITTINGS, WATER, HANDLING & STORAGE							ISDS623	
OIL FIELD, HANDLING							ISDS463	
WATER, HANDLING & STORAGE							ISDS623	
WATER, PROTECTIVE COATING							ISDS613	
PIPELINES								
HOT TAPPING (FLAMMABLE LIQUIDS)				PSD-2201				
PIPING	1926.351							
RADIOGRAPHY							ISDS600	
REPAIR GUIDE (CRUDE OIL, LPG, PRODUCTS)				PSD-2200				
TRANSMISSION, WELDING	1910.252							
WELDING (FLAMMABLE LIQUIDS)				PSD-2201				
PIPING SYSTEMS								
CORROSION PROTECTION	1910.106							
FUEL GASES			B31.2					
GAS - INSTALLATION		54	Z21.30					
(MULTIPLE REFERENCES)			Z83.1					
GAS TRANSMISSION & DISTRIBUTION			B31.8					
HYDROSTATIC TESTS	1910.106							
LIQUID PETROLEUM			B31.4					
MECHANICAL, WELDING	1910.252							
NUCLEAR POWER			B31.7					
PETROLEUM REFINERY			B31.3					
PIPELINE	1926.351							
POWER			B31.1					
PULP MILLS							ISDS243	
REFRIGERATION			B31.5					
SYSTEM IDENTIFICATION			A13.1					
TESTING	1910.106							

APPLICABLE STANDARDS, GUIDES AND PUBLICATIONS

SUBJECT	USDL & USDI	NFPA	ANSI	API	MCA		NSC	MISC
PLANERS								
METAL							ISDS383	
WOOD, POWER FEED							ISDS225	
PLANKS, PLANKING								
EXCAVATIONS	1926.650							
HOISTS & ELEVATORS	1926.552							
SCAFFOLDING	1926.451							
STEEL ERECTION	1926.752							
PLANT (BOTANY)								
SULFUR DIOXIDE				AQM 69-8				
PLANT PROTECTION								
FIREARMS							ISDS413A	
PLASMA-JET TORCH							ISDS552	
PLASTIC FOAM								
MATERIALS HANDLING					SG-5			
STORAGE					SG-5			
PLASTICS								
COMPRESSION MOLDING							ISDS632	
EYE PROTECTION-SPECTRAL-TRANSMISSIVE PROPERTIES			Z2					
PLASTICS INDUSTRY								
AUXILIARY EQUIPMENT	1910.216							
DUST EXPLOSION		654	Z12.16					
EFFECTIVE DATES	1910.216							
(MULTIPLE REFERENCES)	1910.220							
INSTALLATION	1910.216							
MILLS & CALENDERS	1910.216							
STANDARDS SOURCES	1910.221							
PLATERS	1910.261							
PLATFORMS								
BEAM-TYPE	1926.451							
CATCH	1926.451							
CRANES & DERRICKS	1926.550							
ELEVATED			A92.3					
ELEVATING & ROTATING			A92.2					
ELEVATING & ROTATING WORK	1926.451							
GUARDING	1926.500							
LADDER-TYPE (2-PT. SUSPENSION)	1926.451							
LIGHT METAL-TYPE	1926.451							
PILE DRIVING	1926.603							
POWERED - DEFINITION	1910.66							
POWERED - EXTERIOR USE			A120.1					

OCCUPATIONAL SAFETY, HEALTH & FIRE INDEX

SUBJECT	APPLICABLE STANDARDS, GUIDES AND PUBLICATIONS: USDL & USDI	NFPA	ANSI	API	MCA		NSC	MISC
(CON'T)PLATFORMS								
POWERED - GENERAL REQUIREMENTS	1910.66							
POWERED - INSPECTION & TEST	1910.66							
POWERED - PROTECTION OF OPEN-SIDED FLOORS	1910.23							
POWERED - TYPE F	1910.66							
POWERED - TYPE T	1910.66							
TANK CAR					TC-7			
VEHICLE MOUNTED, ELEVATED	1910.67							
PLUMBING UP	1926.752							
PNEUMATIC EQUIPMENT								
DUST EXPLOSION		66	Z12.22					
POWER TOOLS	1926.302							
POWER TOOLS, HAND & PORTABLE	1910.243							
SOUND MEASUREMENT			S5.1					
STEEL ERECTION	1926.752							
TOOL GUARDS	1910.243							
POINT OF OPERATION, SAFEGUARDING	1910.217							
POISON IVY							ISDS304	
POISON OAK							ISDS304	
POISON SUMAC							ISDS304	
POLE HAULING	1926.953							
POLE TOP RESUSCITATION							ISDS376	
POLISHING								
VENTILATION	1910.94		Z43.1					
POLONIUM 210								HG
POLY-1-BUTENE (SEE 1-BUTENE HOMOPOLYMER)								
POLYBUTYLENE (SEE 1-BUTENE HOMOPOLYMER)								
PONTOONS (SEE ALSO HATCH COVERS)								
LIFTING	1918.42							
SECURED	1918.43							
STOWAGE	1918.43							
PORTABLE FIRE EXTINGUISHERS	1910.157							
DISTRIBUTION	1910.157							
GENERAL REQUIREMENTS	1910.157							
INSPECTION, MAINTENANCE & HYDROSTATIC TESTS	1910.157							
SELECTION	1910.157							

SUBJECT	USDL & USDI	NFPA	ANSI	API	MCA		NSC	MISC
	APPLICABLE STANDARDS, GUIDES AND PUBLICATIONS							
PORTABLE GRINDER							ISDS583	
PORTABLE METAL LADDERS	1910.26							
CARE, MAINTENANCE & USE	1910.26							
REQUIREMENTS	1910.26							
TESTING	1910.26							
PORTABLE POWER CHAIN SAW							ISDS320	
PORTABLE REAMER DRILL							ISDS497A	
PORTABLE TOOLS & EQUIPMENT	1910.244							
ABRASIVE BLAST CLEANING NOZZLES	1910.244							
FRAYED CABLES	1915.72							
(MULTIPLE REFERENCES)	1916.72							
(MULTIPLE REFERENCES)	1917.72							
GROUNDING	1915.23							
(MULTIPLE REFERENCES)	1915.72							
(MULTIPLE REFERENCES)	1916.23							
(MULTIPLE REFERENCES)	1916.72							
(MULTIPLE REFERENCES)	1917.72							
GUARDS	1915.23							
(MULTIPLE REFERENCES)	1916.23							
JACKS	1910.244							
MECHANICAL PAINT REMOVERS	1915.23							
(MULTIPLE REFERENCES)	1916.23							
NOTIFYING OFFICER	1915.72							
(MULTIPLE REFERENCES)	1916.72							
POWER TOOLS	1910.241							
SWITCHES	1915.72							
(MULTIPLE REFERENCES)	1916.72							
(MULTIPLE REFERENCES)	1917.72							
WOODWORKING GUARD	1910.243							
PORTABLE WOOD LADDERS	1910.25							
APPLICATION OF REQUIREMENTS	1910.25							
CARE & USE	1910.25							
CONSTRUCTION REQUIREMENTS	1910.25							
MATERIALS	1910.25							
PORTLAND CEMENT								
CONCRETE PAVING							ISDS541	
POTABLE WATER	1910.141							
POTASSIUM DICHROMATE					SD-46			

SUBJECT	APPLICABLE STANDARDS, GUIDES AND PUBLICATIONS: USDL & USDI	NFPA	ANSI	API	MCA		NSC	MISC
POTASSIUM HYDROXIDE (SEE CAUSTIC POTASH)								
POWDER ACTUATED FASTENING TOOLS	1915.75							
(MULTIPLE REFERENCES)	1916.75		A10.3					
OPERATOR INSTRUCTIONS	1915.75							
(MULTIPLE REFERENCES)	1916.75							
PRECAUTIONS	1915.75							
(MULTIPLE REFERENCES)	1916.75							
POWDER ACTUATED HAND TOOLS			A10.3				ISDS236	
GUARDING			A10.3					
POWDER COATINGS, SPRAY FINISHING	1910.107							
CLEANING	1910.107							
ELECTRIC WIRING	1910.107							
POWER CHAIN SAW								
MAINTENANCE	1910.107							
PORTABLE							ISDS320	
SMOKING	1910.107							
POWER HAND TRUCK							ISDS317	
POWER LAWN MOWER							ISDS464	
GUARDING			B71.1					
ROTARY	1910.243		B71.1					
POWER LINES, CRANES & DERRICKS	1926.550							
POWER PLATFORMS FOR EXTERIOR BUILDING MAINTENANCE	1910.66							
CLEANING	1910.66							
DEFINITION	1910.66							
GENERAL REQUIREMENTS	1910.66							
INSPECTION & TEST	1910.66							
MAINTENANCE	1910.66							
RECORDS	1910.66							
TESTING	1910.66							
TYPE F	1910.66							
TYPE T	1910.66							
POWER PRESS								
AIR CONTROLLING EQUIPMENT	1910.217							
CONSTRUCTION, CARE, USE			B11.1					
DIES							ISDS211	
ELECTRICAL CONTROLS							ISDS624	
HANDLING FINISHED PIECES							ISDS534	
MECHANICAL INSPECTION & MAINTENANCE							ISDS603	
RECORDS	1910.217							

SUBJECT	USDL & USDI	NFPA	ANSI	API	MCA		NSC	MISC
	APPLICABLE STANDARDS, GUIDES AND PUBLICATIONS							
POWER SHOVELS								
OPERATION OF							ISDS271	
POWER TOOLS, HAND & PORTABLE	1926.302							
ABRASIVE WHEELS	1910.243							
COMPRESSED AIR CLEANING	1910.242							
DEFINITION	1910.241							
EFFECTIVE DATES	1910.245							
EMPLOYEE	1910.242							
EXPLOSIVE ACTUATED FASTENINGS	1910.243							
GUARDING	1910.243							
LAWN MOWERS, POWER	1910.243							
PNEUMATIC POWERED	1910.243							
STANDARDS SOURCES	1910.246							
WOODWORKING	1910.243							
POWER TRANSMISSION & DISTRIBUTION LINES	1926.950							
CLEARANCE	1926.950							
DEENERGIZING LINES & EQUIPMENT	1926.950							
EMERGENCY PROCEDURES	1926.950							
EMPLOYEE INSULATION	1926.950							
ENERGIZED PARTS	1926.950							
EXISTING LINES & EQUIPMENT	1926.950							
FIRST AID	1926.950							
HOT STICK DISTANCES	1926.950							
HYDRAULIC FLUIDS	1926.950							
MECHANICAL - SAFETY CODE			B15.1					
MINIMUM DISTANCES	1926.950							
NIGHT WORK	1926.950							
SANITATION	1926.950							
WATER	1926.950							
WORKING DISTANCES	1926.950							
POWER TRANSMISSION APPARATUS, MECHANICAL	1910.219							
APPROVED MATERIALS	1910.219							
BELTS, ROPE & CHAIN DRIVES	1910.219							
BELTS,SHIFTERS,CLUTCHES,POLES,PERCHES & SHIPPERS	1910.219							
CARE	1910.219							
CLUTCHES	1910.217							
COLLARS & COUPLINGS	1910.219							
DISKS, SHIELDS & "U" GUARDS	1910.219							
GEARS, SPROCKETS & CHAINS	1910.219							
GENERAL REQUIREMENTS	1910.219							
GUARDING CLUTCHES, CUT-OFF COUPLING & CLUTCH PUL	1910.219							
GUARDING FRICTION DRIVES	1910.219							
KEYS, SET SCREWS & OTHER PROJECTIONS	1910.219							
PRIME-MOVER GUARDS	1910.219							
PULLEYS	1910.219							
SHAFTING	1910.219							

SUBJECT	USDL & USDI	NFPA	ANSI	API	MCA		NSC	MISC
	APPLICABLE STANDARDS, GUIDES AND PUBLICATIONS							
POWER WINDLASS MACHINE	1910.214							
PRESSES								
FORGING	1910.218							
GRAPHIC ARTS			B65.1					
INSPECTION & MAINTENANCE	1910.217							
KICK (FOOT)							ISDS363	
MECHANICAL FORGING							ISDS560	
OPERATION	1910.217							
TERMS	1910.211							
TRIMMING	1910.218							
PRESSES, BRAKES							ISDS419A	
PRESSES, GUARDING	1910.217							
PRESSES, POWER, MECHANICAL	1910.217							
DESIGN, CONSTRUCTION, SETTING & FEEDING OF DIES	1910.217							
GENERAL REQUIREMENTS	1910.217							
INSPECTION, MAINTENANCE & MODIFICATION	1910.217							
MECHANICAL POWER PRESS GUARDING & CONSTRUCTION	1910.217							
OPERATION	1910.217							
SAFEGUARDING POINT OF OPERATION	1910.217							
PRESSURE REGULATOR								
LPG - SAFETY STANDARDS			Z217.1					UL 144
PRESSURE VESSELS	1910.106							
(MULTIPLE REFERENCES)	1926.29							
AIR RECEIVERS	1910.169							
DESIGN & CONSTRUCTION	1915.101							
(MULTIPLE REFERENCES)	1916.101							
INSPECTION	1915.101							
(MULTIPLE REFERENCES)	1916.101							
PILE DRIVING EQUIPMENT	1926.603							
PULP MILLS							ISDS243	
SAFE WORKING PRESSURE	1915.101							
(MULTIPLE REFERENCES)	1916.101							
PREVENTER								
COAMING ROLLERS	1918.52							
HEEL BLOCKS	1918.52							
PREVENTER GUYS								
SECURING	1918.54							
STRENGTH	1918.52							
TAILS	1918.52							

SUBJECT	USDL & USDI	NFPA	ANSI	API	MCA		NSC	MISC
	APPLICABLE STANDARDS, GUIDES AND PUBLICATIONS							
PRIME-MOVER, GUARDS (FLYWHEELS)	1910.219							
PRIVY, SPECIFICATIONS	1910.143							
PROCESS OPERATOR								
TRAINING					SG-15			
PROCESSING PLANTS								
DRAINAGE	1910.106							
ELECTRIC WIRING	1910.106							
FIRE PROTECTION	1910.106							
FLAMMABLE & COMBUSTIBLE LIQUIDS	1910.106							
LOADING	1910.106							
MAINTENANCE	1910.106							
PROJECTIONS	1910.219							
PROOF TESTS								
AFTER REPAIRS	1919.34							
GEAR, LOOSE	1919.31							
ORDER OF	1919.35							
ROPE, WIRE	1919.33							
PROPANOL (SEE PROPYL ALCOHOL)								
PROPELLANT								
SMOKELESS	1910.109							
PROPENE					SD-59			
BETA-PROPIOLACTONE	1910.93							
PROPIONIC ACID					CIC-71			
PROPYL ALCOHOL								HG
PROPYLENE (SEE PROPENE)								
PROPYLENE DICHLORIDE								HG
PROPYLENE GLYCOL (SEE 1,2-PROPANEDIOL)								
PROPYLENE OXIDE					CIC-72			HG
PROTECTIVE EQUIPMENT								
ABRASIVE BLASTERS	1915.23							
(MULTIPLE REFERENCES)	1916.23							
ABRASIVE BLASTING	1910.94							
ACID & CHEMICAL TANKS	1910.94							
ADDITIONAL DELAY IN EFFECTIVE DATE	1910.138							

SUBJECT	USDL & USDI	NFPA	ANSI	API	MCA		NSC	MISC
		APPLICABLE STANDARDS, GUIDES AND PUBLICATIONS						
(CON'T)PROTECTIVE EQUIPMENT								
APPLICATION	1910.132							
ASBESTOS	1910.93							
CANOPY	1910.266							
CLEANING PETROLEUM STORAGE TANK				RP-2015				
COAL DUST - RESPIRATORY PROTECTION			Z88.4					
CUTTING, WELDING, HEATING	1926.353							
DESIGN	1910.132							
DROWNING PROTECTION	1918.106							
EAR PROTECTION	1926.101		Z24.22					
EFFECTIVE DATES	1910.138							
ELECTRICAL PROTECTIVE DEVICES	1910.137							
(MULTIPLE REFERENCES)	1910.265							
ELECTRICAL SHOCK	1910.137							
ELECTRICAL WORKER			J6					
ELECTRICAL WORKER - FLEXIBLE INSULATED							ISDS598	
ELECTRICAL WORKER - RUBBER INSULATING BLANKET			J6.4					ASTM D1048-59
ELECTRICAL WORKER - RUBBER INSULATING GLOVES			J6.6					ASTM D120-68
ELECTRICAL WORKER - RUBBER INSULATING HOOD			J6.2					ASTM D1049-59
ELECTRICAL WORKER - RUBBER INSULATING HOSE			J6.1					ASTM D1050-59
ELECTRICAL WORKER - RUBBER INSULATING SLEEVES	1910.137		J6.5					ASTM D1051-59
ELECTRICAL WORKER - RUBBER MATTING			J6.7					ASTM D178-24
EMERGENCY SHOWERS & FOUNTAINS, OPEN SURFACE TANK	1910.94							
EMERGENCY SHOWERS & FOUNTAINS, PULP, PAPER MILLS	1910.261							
EMPLOYEE	1910.132							
EMPLOYEE OWNED	1910.132							
EMPLOYER FURNISHED	1910.133							
EXCAVATIONS	1926.650							
EYE & FACE PROTECTION	1910.133							
EYE PROTECTION	1910.133							
(MULTIPLE REFERENCES)	1918.101							
(MULTIPLE REFERENCES)	1926.102		Z87.1					
EYE PROTECTION-SPECTRAL-TRANSM PROPS OF PLASTICS			Z2					
FACE PROTECTION			Z87.1					
FACE PROTECTION - WELDING	1910.252							
FEDERAL SUPPLY CONTRACTS	50-204.7							
FOOT PROTECTION	1910.136							
(MULTIPLE REFERENCES)	1918.104		Z41.1					
FOOT PROTECTION, SHOES (BLASTING)	1910.94							
GAS MASK CANISTER IDENTIFICATION			K13.1					
GENERAL REQUIREMENTS	1910.132							
GLOVES	1915.83							
(MULTIPLE REFERENCES)	1916.83							
(MULTIPLE REFERENCES)	1917.83							
(MULTIPLE REFERENCES)	1926.951							
GLOVES, APRONS	1910.94							
GOGGLES	1910.133							
GOGGLES, LASERS	1926.102							
GOGGLES, WELDING	1926.102							
GREASY CLOTHING	1915.83							
(MULTIPLE REFERENCES)	1916.83							
(MULTIPLE REFERENCES)	1917.83							

APPLICABLE STANDARDS, GUIDES AND PUBLICATIONS

SUBJECT	USDL & USDI	NFPA	ANSI	API	MCA		NSC	MISC
(CON'T)PROTECTIVE EQUIPMENT								
HARDHATS	1910.135							
HATS	1926.951							
HEAD PROTECTION	1910.135							
(MULTIPLE REFERENCES)	1915.83							
(MULTIPLE REFERENCES)	1916.83							
(MULTIPLE REFERENCES)	1917.83							
(MULTIPLE REFERENCES)	1918.105							
(MULTIPLE REFERENCES)	1926.100		Z89.1					
HEADGEAR - VEHICULAR USER			Z90.1					
HEARING PROTECTION	1910.95							
(MULTIPLE REFERENCES)	1926.101							
HELMET FOR ELECTRICAL WORKERS			Z89.2					
LIQUEFIED		508						
MARKINGS	1910.133							
NOISE EXPOSURE	1910.95							
NONRUBBER	1926.951							
OCCUPATIONAL FOOT PROTECTION	1910.136							
OPEN-PIT MINE	55.15							
OPEN-SURFACE TANKS	1910.94							
OVERHEAD PROTECTION	1926.451							
(MULTIPLE REFERENCES)	1926.1003							
PAINTING	1915.24							
(MULTIPLE REFERENCES)	1916.24							
PERSONAL	1926.28							
(MULTIPLE REFERENCES)	1926.100							
(MULTIPLE REFERENCES)	1926.551							
(MULTIPLE REFERENCES)	1926.951							
PERSONAL (WELDING)	1910.252							
PROTECTIVE CLOTHING	1910.94							
(MULTIPLE REFERENCES)	1918.103							
PULP & PAPER MILLS	1910.261							
PULPWOOD LOGGING	1910.266							
RADIATION	1910.96							
RADIATION - ELECTROMAGNETIC	1910.97							
RADIATION - IONIZING	1910.96							
RADIATION - NUCLEAR FUEL PLANT			N7.2					
RADIATION - URANIUM MINES			N7.1					
RECORDS - FIRE RESISTANCE CLASSIFICATION			Z134.1					UL 72
RESPIRATORS	1910.93							
(MULTIPLE REFERENCES)	1910.94							
RESPIRATORY							ISDS444A	
RESPIRATORY EQUIPMENT - DUST	70.300							
RESPIRATORY PROTECTION	1910.134							
(MULTIPLE REFERENCES)	1918.102		Z88.2					
RESPIRATORY PROTECTION - RADON DAUGHTERS			Z88.1					
RUBBER	1910.137							
(MULTIPLE REFERENCES)	1926.951							
RUBBER PROTECTIVE EQUIPMENT	1910.137							
SAFETY BELTS	1926.104							

SUBJECT	USDL & USDI	NFPA	ANSI	API	MCA		NSC	MISC
	APPLICABLE STANDARDS, GUIDES AND PUBLICATIONS							
(CON'T)PROTECTIVE EQUIPMENT								
SAFETY BELTS, MATERIALS STORAGE	1926.250							
SAFETY BELTS, PILE DRIVING	1926.603							
SAFETY BELTS, SCAFFOLDING	1926.451							
SAFETY BELTS, STEEL ERECTION	1926.752							
SAFETY HAT							ISDS561	
SAFETY NET	1926.105							
SAFETY NET, CONSTRUCTION							ISDS608	
SAFETY NET, STEEL ERECTION	1926.750							
SAFETY SHOES			Z41.1					
SAND, GRAVEL, CRUSHED STONE OPERATION	56.15							
SAWMILLS	1910.265							
STANDARDS SOURCES	1910.139							
(MULTIPLE REFERENCES)	1910.141							
STORAGE	1910.107							
TEXTILES	1910.262							
TOXIC CLEANING SOLVENTS	1915.21							
(MULTIPLE REFERENCES)	1916.21							
UNDERGROUND MINE	57.15							
UNIT FIRST-AID KIT							ISDS202	
WELDING	1910.252							
WET ENVIRONMENT	1910.94							
PUBLIC RELATIONS								
EMERGENCY					SG-12			
PULLEYS, POWER TRANSMISSION	1910.219							
PULP, PAPER & PAPERBOARD MILLS								
BARKING DEVICES	1910.261							
BELT CONVEYORS	1910.261							
BLEACHING	1910.261							
BRIDGE OR DOCK PLATES	1910.261							
CHEMICAL PROCESSES	1910.261							
CRANES	1910.261							
EFFECTIVE DATES	1910.261							
FINISHING ROOMS	1910.261							
FLOORS	1910.261							
GUARDS	1910.261							
HAND TOOLS	1910.261							
HANDLING	1910.261							
LIGHTING	1910.261							
LOCKOUTS	1910.261							
MACHINE ROOMS	1910.261							
MATERIALS HANDLING	1910.261							
MECHANICAL PULP PROCESSES	1910.261							
PERSONAL PROTECTIVE CLOTHING	1910.261							
PERSONAL PROTECTIVE EQUIPMENT	1910.261							
PLATFORMS	1910.261							
PULPWOOD PREPARATION	1910.261							

APPLICABLE STANDARDS, GUIDES AND PUBLICATIONS

SUBJECT	USDL & USDI	NFPA	ANSI	API	MCA		NSC	MISC
(CON'T)PULP, PAPER & PAPERBOARD MILLS								
PULPWOOD REMOVAL	1910.261							
RAGS AND OLD PAPER	1910.261							
SAFE PRACTICES	1910.261							
SIGNS, CONVEYORS	1910.261							
SIGNS, TRAFFIC	1910.261							
STANDARDS SOURCES	1910.261							
STOCK PREPARATION	1910.261							
STORAGE	1910.261							
TANKS	1910.261							
TRAFFIC WARNING SIGNS	1910.261							
TRUCK, INDUSTRIAL POWERED	1910.261							
PULVERIZED - FUEL SYSTEM								
DUST EXPLOSION		60	Z12.1					
PUMPS								
LOADING	1910.106							
PURGED ENCLOSURE								
ELECTRICAL EQUIPMENT		496	C106.1					
PYRIDINE							ISDS310A	HG
PYROXYLIN PLASTIC								
IN FACTORIES		42						
SALE		43						
STORAGE		43						

SUBJECT	APPLICABLE STANDARDS, GUIDES AND PUBLICATIONS							
	USDL & USDI	NFPA	ANSI	API	MCA		NSC	MISC
QUINONE (SEE BENZOQUINONE)								

SUBJECT	USDL & USDI	NFPA	ANSI	API	MCA		NSC	MISC
	APPLICABLE STANDARDS, GUIDES AND PUBLICATIONS							
RACEWAYS								
SURFACE METAL RACEWAYS & FITTINGS			C33.50					UL 5
RADAR	1915.55							
(MULTIPLE REFERENCES)	1916.55							
HAZARDS							ISDS481	
RADIAL SAWS	1910.213							
RADIANT ENERGY PROTECTION								
LASERS	1926.102							
WELDING	1926.102							
RADIANT HEAT CONTROL							ISDS381A	
RADIATION (SEE ALSO IONIZING RADIATION)								
ALARM TESTING	1910.96							
AREAS, POSTING INSTRUCTIONS	1910.96							
CAUTION SIGNS	1910.96							
DOSIMETRY FOR CRITICALITY ACCIDENTS			N13.3					
ELECTROMAGNETIC	1910.97		C95.1					
EMPLOYEE RECORDS, DISCLOSURE	1910.96							
EMPLOYER RECORDS	1910.96							
EVACUATION SIGNAL	1910.96							
EXCEPTIONS	1910.96							
EXPOSURE	1910.96							
(MULTIPLE REFERENCES)	50-204.21							
EXPOSURE EVACUATION SIGNAL			N2.3					
EXPOSURE RECORD			N13.6					
EXPOSURE SIGNAL			N2.3					
FEDERAL SUPPLY CONTRACTS	50-204.20							
FLUORESCENCE ANALYSIS			N43.2					
GAMMA-RAY - PORTABLE SURVEY INSTRUMENT			N13.4					
GAMMA-RAY DOSIMETER			N13.5					
HAZARDS, PROCEDURES PRECAUTIONARY	1910.96							
INCIDENT NOTIFICATION	1910.96							
(MULTIPLE REFERENCES)	50-204.30							
INERT-GAS METAL-ARC WELDING	1926.353							
IONIZING	1910.96							
(MULTIPLE REFERENCES)	1926.53							
MONITORING - ADMINISTRATIVE PRACTICE			N13.2					
NONIONIZING	1910.97							
(MULTIPLE REFERENCES)	1926.54							
NUCLEAR FUEL PLANT PROTECTION			N7.2					
NUCLEAR POWER GENERATING STATION PROTECTION			N42.7					IEEE 279
PACKAGING & MARKING EXEMPTIONS	1910.96							
PARTICLE ACCELERATOR SAFETY			N43.1					NBS H107
PERSONAL MONITORING	1910.96							
PERSONNEL POSTING	50-204.27							
PRECAUTIONS & MONITORING	50-204.23							

APPLICABLE STANDARDS, GUIDES AND PUBLICATIONS

SUBJECT	USDL & USDI	NFPA	ANSI	API	MCA		NSC	MISC
(CON'T)RADIATION (SEE ALSO IONIZING RADIATION)								
RADIO FREQUENCY - HAZARD WARNING			C95.2					
RECORDS	1910.96							
(MULTIPLE REFERENCES)	50-204.32							
REPORTS OF OVEREXPOSURE & EXCESSIVE LEVELS	1910.96							
RESPIRATORY PROTECTION - RADON DAUGHTERS			Z88.1					
RESTRICTED AREAS	1910.96							
SIGNS, LABELS, SIGNALS	50-204.24							
STANDARDS FOR MINING	1910.96							
STORAGE	1910.96							
SYMBOL			N2.1					ISO R361
UNDERGROUND MINE	57.5							
URANIUM MINES PROTECTION			N13.8					
(MULTIPLE REFERENCES)			N7.1					
WASTE	1910.96							
WASTE DISPOSAL	50-204.29							
X-RAY - PORTABLE SURVEY INSTRUMENT			N13.4					
X-RAY DIFFRACTION			N43.2					
X-RAY DOSIMETER			N13.5					
RADIO								
SAFETY STANDARD			C33.55					UL 492
RADIO FREQUENCY HEATING							ISDS319	
RADIOACTIVE								
AIRBORNE, EXPOSURE LIMITS	1910.96							
DOSIMETRY FOR CRITICALITY ACCIDENTS			N13.3					
FACILITIES HANDLING - RADIOACTIVE MATERIALS		801						
MATERIALS EXPOSURE	1910.96							
MATERIALS STORAGE	50-204.28							
MATERIALS, INCIDENT MODIFICATION	1910.96							
MATERIALS, PACKAGING & EXEMPTIONS	1910.96							
MATERIALS, REPORT OF EXCESSIVE LVLS OF CONC	1910.96							
MATERIALS, REPORT OF OVEREXPOSURE	1910.96							
RADIOACTIVE MATERIALS WASTE	1910.96							
RESPIRATORY PROTECTION - RADON DAUGHTERS			Z88.1					
SAMPLING AIRBORNE MATERIAL			N13.1					
RADIOGRAPHY								
PIPELINE							ISDS600	
RADIOLOGY								
PARTICLE ACCELERATOR SAFETY			N43.1					
RADON & ITS DAUGHTERS								HG

SUBJECT	APPLICABLE STANDARDS, GUIDES AND PUBLICATIONS: USDL & USDI	NFPA	ANSI	API	MCA		NSC	MISC
RAIL SWEEPS, OVERHEAD GANTRY CRANES	1910.179							
RAILINGS (SEE GUARDRAILS)								
RAILROAD								
HIGHWAY GRADE CROSSING			D8.1					AAR BULL 6
RAILROAD CARS	1918.84							
INDUSTRIAL TRUCKS IN	1910.178							
MATERIALS HANDLING & STORAGE	1910.176							
ROLLING	1910.176							
STOPS, POSITIVE BRAKING	1910.178							
RAMPS	1918.23							
(MULTIPLE REFERENCES)	1918.24							
ENCLOSURES	1910.23							
EXIT	1910.37							
RATED CAPACITY								
CHAIN SLINGS	1926.251							
SYNTHETIC WEBBING	1926.251							
RATING CAPACITY	1926.252							
REACTOR								
SAFE WORKING PRECAUTIONS				PSD-2211				
REAMER DRILL								
PORTABLE							ISDS497A	
RECEIVERS & TRANSMITTERS								
TOWER, TRANSMITTER	1926.550							
RECORDS								
ACCIDENT							ISDS527A	
ACCREDITED PERSONS	1919.11							
ARCHIVES PROTECTION		232AM						
ASBESTOS	1910.93							
BUILDING INSPECTION	1910.66							
COMMUNICABLE DISEASES	1910.142							
CRANES, LOCOMOTIVE, CRAWLER & TRUCK	1910.180							
CRANES, OVERHEAD & GANTRY	1910.179							
DERRICKS	1910.181							
FIRE EXTINGUISHERS, FIXED DRY CHEMICAL	1910.160							
FIRE EXTINGUISHERS, PORTABLE	1910.157							
FORGING EQUIPMENT INSPECTION	1910.218							
INJURY REPORTING, WELDING	1910.252							
IONIZING RADIATION	1910.96							
LABOR CAMPS	1910.142							
LIBRARY COLLECTION PROTECTION		910						

SUBJECT	USDL & USDI	NFPA	ANSI	API	MCA		NSC	MISC
	APPLICABLE STANDARDS, GUIDES AND PUBLICATIONS							
(CON'T)RECORDS								
LIQUID STORAGE TANKS, CLASS I	1910.106							
MANLIFTS	1910.68							
MECHANICAL POWER PRESSES	1910.217							
MUSEUM COLLECTION PROTECTION		911						
OFF-THE-JOB							ISDS601	
PERSONAL MONITORING, ASBESTOS	1910.93							
PERSONAL MONITORING, IONIZING RADIATION	1910.96							
POWER PLATFORMS INSPECTION	1910.66							
POWER PRESS INSPECTION	1910.217							
PROTECTION		232						
PROTECTION EQUIPMENT - FIRE RESISTANCE CLASS			Z134.1					
RADIATION EXPOSURE	1910.96							
(MULTIPLE REFERENCES)	50-204.32		N13.6					
RECORDS, DISCLOSURE, IONIZING RADIATION	1910.96							
RESPIRATORS	1910.134							
VESSELS	1919.12							
WELDING	1910.252							
REEFER COMPARTMENTS, SECURING	1918.83							
REFINERIES	1910.106							
FIRE PROTECTION	1910.106							
WHARVES	1910.106							
REFLECTING MATERIAL							ISDS245	
REFRIGERATION								
DRINKING WATER COOLER			A112.11.1					ARI 1010
HOME - SAFETY STANDARD			B97.1					UL 250
MECHANICAL - SAFETY CODE			B9.1					ASHRAE 15
PIPING			B31.5					
SYSTEMS	1910.177							
REFUELING								
CRANES	1910.180							
DERRICKS	1910.181							
TRUCK	1910.178							
REFUSE DISPOSAL, LABOR CAMPS	1910.142							
REGULATORS	1910.252							
(MULTIPLE REFERENCES)	1926.350							
REINFORCING STEEL	1926.700							

SUBJECT	USDL & USDI	NFPA	ANSI	API	MCA		NSC	MISC
	APPLICABLE STANDARDS, GUIDES AND PUBLICATIONS							
RELAY								
DIRECTION	1910.66							
POTENTIAL FOR VERTICAL TRAVEL	1910.66							
REMOTE GAS PUMPING SYSTEMS	1910.106							
REPAIR OPERATIONS, MEANS OF EGRESS	1910.36							
REPORTING								
BENZIDINE INCIDENTS OR OPERATIONS	1910.93							
COMMUNICABLE DISEASES	1910.142							
FIRE - CODING SYSTEM		901						
FIRE - FIELD INCIDENT MANUAL		901A						
UNDERWATER ACCIDENT REPORT FORM			Z86.2					
RESIDUE DISPOSAL (SEE WASTE DISPOSAL)								
RESPIRATORS							ISDS444A	
AIR QUALITY	1910.134							
CARE & MAINTENANCE	1910.134							
FIRE FIGHTER		19B						
IDENTIFICATION OF GAS MASK CANISTERS	1910.134							
LIMITS	1910.94							
MAINTENANCE	1910.134							
MARKINGS	1910.134							
MINIMAL PROGRAM ACCEPTABLE	1910.134							
PROGRAM, ASBESTOS	1910.93							
PROTECTION	1910.134		Z88.2					
PROTECTIVE - RADON DAUGHTERS			Z88.1					
PROTECTIVE EQUIP	1910.93							
RECORDS	1910.134							
SELECTION	1910.134							
TRAINING PERSONNEL	1910.134							
RESPIRATORY PROTECTION	1926.103							
ABRASIVE BLASTERS	1915.23							
(MULTIPLE REFERENCES)	1916.23							
ATMOSPHERES, LIFE HAZARDS	1915.82							
(MULTIPLE REFERENCES)	1916.82							
(MULTIPLE REFERENCES)	1917.82							
CANISTERS, GAS MASK	1918.102							
CARBON MONOXIDE	1918.93							
CARTRIDGE RESPIRATORS	1918.102							
CLEANED & DISINFECTED	1918.102							
COAL MINE DUST			Z88.4					
CONDITIONS	1918.93							
CONTAMINANTS	1915.82							
(MULTIPLE REFERENCES)	1916.82							
(MULTIPLE REFERENCES)	1917.82							
DUST	1918.93							
(MULTIPLE REFERENCES)	70.300							

SUBJECT	USDL & USDI	NFPA	ANSI	API	MCA		NSC	MISC
	APPLICABLE STANDARDS, GUIDES AND PUBLICATIONS							
(CON'T)RESPIRATORY PROTECTION								
EXCAVATIONS	1926.650							
EXCEPTIONS	1915.11							
(MULTIPLE REFERENCES)	1917.11							
FIRE SERVICE			Z88.5					
FLAMMABLE	1915.11							
(MULTIPLE REFERENCES)	1917.11							
FUMIGANTS	1918.93							
GAS FREE	1915.14							
(MULTIPLE REFERENCES)	1917.14							
GAS MASK	1918.102							
GASEOUS CONTAMINANTS	1915.82							
(MULTIPLE REFERENCES)	1916.82							
(MULTIPLE REFERENCES)	1917.82							
GENERAL	1915.82							
(MULTIPLE REFERENCES)	1916.82							
(MULTIPLE REFERENCES)	1917.82							
HOME THERAPY		56HM						
INSPECTION & MAINTENANCE	1918.102							
INSTRUCTIONS	1918.102							
LIFE HAZARDS	1918.93							
MECHANICAL PAINT REMOVERS	1915.23							
(MULTIPLE REFERENCES)	1916.23							
MICROFLORA & RODENTS				MRR-7101				
OXYGEN DEFICIENT	1915.11							
(MULTIPLE REFERENCES)	1917.11							
PAINTING	1915.24							
(MULTIPLE REFERENCES)	1916.24							
PARTICULATE CONTAMINANTS	1917.82							
PARTICULATES	1915.82							
(MULTIPLE REFERENCES)	1916.82							
POWDER ACTUATED FASTENING TOOLS	1915.75							
(MULTIPLE REFERENCES)	1916.75							
REQUIREMENTS	1918.102							
RESPIRATORS	1926.103							
RESPIRATORS, CHEMICAL CARTRIDGE	1918.102							
RESPIRATORS, DUST	1918.102							
TOXIC	1915.11							
(MULTIPLE REFERENCES)	1917.11							
TOXIC CLEANING SOLVENTS	1915.21							
(MULTIPLE REFERENCES)	1916.21							
TOXIC PRESERVATIVE COATINGS	1915.33							
(MULTIPLE REFERENCES)	1916.33							
(MULTIPLE REFERENCES)	1917.33							
WELDING	1915.31							
(MULTIPLE REFERENCES)	1916.31							
(MULTIPLE REFERENCES)	1917.31							
WELDING & CUTTING	1926.353							

SUBJECT	USDL & USDI	NFPA	ANSI	API	MCA		NSC	MISC
	APPLICABLE STANDARDS, GUIDES AND PUBLICATIONS							
REST ROOMS, WOMEN	1910.141							
RESUSCITATION								
POLE TOP							ISDS376	
REWINDERS	1910.261							
RIGGING								
CRANES & DERRICKS	1926.550							
HOISTS & ELEVATORS	1926.550							
MATERIALS HANDLING	1926.251							
RIGGING GEAR								
BOOM PLACEMENT	1918.54							
FALLS, CHAFING	1918.54							
GUY PLACEMENT	1918.54							
PREVENTER SECURING	1918.54							
RING BUOYS	1926.106							
RIPSAW, HAND-FED	1910.213							
COOPERAGE MACHINERY	1910.214							
RIVETING	1926.752							
RODENT CONTROL, LABOR CAMPS	1910.142							
ROLLERS								
BUILDING FACE	1910.66							
GUIDE	1910.66							
ROLLER CONVEYORS							ISDS528	
ROLLOVER PROTECTIVE STRUCTURES (SEE ROPS)								
ROOF CAR	1910.66							
ROOF COVERINGS								
FIRE TESTS		256						
ROOF JACK								
MOBILE HOME			A160.1					UL 311
ROOFING BRACKETS	1910.28							
(MULTIPLE REFERENCES)	1926.451							

SUBJECT	USDL & USDI	NFPA	ANSI	API	MCA		NSC	MISC
	APPLICABLE STANDARDS, GUIDES AND PUBLICATIONS							
ROOM							ISDS423	
WOOD								
ROPE								
CRANES	1910.179							
(MULTIPLE REFERENCES)	1910.180							
CRANES & DERRICKS	1926.550							
DERRICKS	1910.181							
EYE SPLICES	1919.24							
MANILA ROPE & SLINGS	1915.62							
(MULTIPLE REFERENCES)	1916.62							
(MULTIPLE REFERENCES)	1917.62							
MANILA, SAFE WORKING LOADS	1915.68							
(MULTIPLE REFERENCES)	1916.68							
(MULTIPLE REFERENCES)	1917.68							
NATURAL & SYNTHETIC	1926.251							
WIRE ROPE & SLINGS	1915.62							
(MULTIPLE REFERENCES)	1916.62							
(MULTIPLE REFERENCES)	1917.62							
WIRE, SAFE WORKING LOADS	1915.68							
(MULTIPLE REFERENCES)	1916.68							
(MULTIPLE REFERENCES)	1917.68							
ROPE GUARDS, OVERHEAD GANTRY CRANES	1910.179							
ROPE INSPECTION								
CRANES, CRAWLER & TRUCK	1910.179							
DERRICKS	1910.181							
OVERHEAD GANTRY CRANES	1910.179							
ROPE RATED LOAD	1910.66							
TAG DATE	1910.66							
ROPS (ROLLOVER PROTECTIVE STRUCTURES)								
COVERAGE	1926.1000							
CRAWLER TRACTORS	1926.1001							
DOZERS	1926.1001							
EFFECTIVE DATES	1926.1000							
GRADERS	1926.1001							
LABELING	1926.1000							
LOADERS	1926.1001							
PERFORMANCE REQUIREMENTS	1926.1001							
(MULTIPLE REFERENCES)	1926.1002						ISDS622	
PROTECTIVE FRAMES	1926.1002							
REMOUNTING	1926.1000							
SCRAPERS	1926.1001							
STANDARDS SOURCES	1926.1001							
TEST PROCEDURES	1926.1001							
(MULTIPLE REFERENCES)	1926.1002							
TEST SETUPS	1926.1001							

APPLICABLE STANDARDS, GUIDES AND PUBLICATIONS

SUBJECT	USDL & USDI	NFPA	ANSI	API	MCA		NSC	MISC
(CON'T)ROPS (ROLLOVER PROTECTIVE STRUCTURES)								
TRACTORS	1910.178							
ROSETTES								
SAFETY STANDARD			C33.20					UL 351
ROTARY								
CUTTERS	1910.261							
GUARDING	1910.219							
LAWN MOWERS	1910.243							
RUBBER								
TANK & VESSEL LINING							ISDS492A	
RUBBER GLOVES, BLANKETS	1910.137							
RUBBER PROTECTIVE EQUIPMENT, ELECTRICAL SHOCK	1910.137							
RUNNERS (SEE FALLS (RUNNERS))								
RUNWAYS								
EXCAVATIONS	1926.650							
GUARDING	1926.500							
PROTECTION PLATFORMS	1910.23							
RUST REMOVERS, CHEMICAL	1910.94							

SUBJECT	USDL & USDI	NFPA	ANSI	API	MCA		NSC	MISC
			APPLICABLE STANDARDS, GUIDES AND PUBLICATIONS					
SAFE-ENDING								
IMPACT TOOL							ISDS414	
SAFE WORKING LOADS	1926.252							
ABSENCE OF DATA	1919.75							
CARGO GEAR	1919.13							
CRANES	1919.75							
(MULTIPLE REFERENCES)	1919.76							
DERRICKS	1919.73							
(MULTIPLE REFERENCES)	1919.75							
(MULTIPLE REFERENCES)	1919.76							
(MULTIPLE REFERENCES)	1919.77							
RATINGS	1919.73							
ROPE, WIRE	1919.33							
SAFETY								
AIR DRILLING				RP-2011				
API FIRST AID INSTRUCTOR TRAINING GUIDE				BULL-2019				
CHEMICAL PROCESSES					SG-14			
CHEMICAL REFERENCES							ISDS486A	
COLOR CODES	1910.144		Z53.1					
CONTROLS, CALENDERS	1910.216							
CONTROLS, MILLS	1910.216							
DIVING-OFFSHORE PRODUCTION				TB-7104				
DRILLING				RP-2010				
DRIVER IMPROVEMENT COURSE				RP-2020				
ELECTRIC CUTTING EQUIPMENT - PETROLEUM OPERATION				RP-2009				
ELECTRIC WELDING EQUIPMENT - PETROLEUM OPERATION				RP-2009				
EQUIPMENT MODIFICATION & PROCUREMENT	1910.132							
FALLS ON FLOOR							ISDS495	
FEDERAL SUPPLY CONTRACTS	50-204.1							
FIRE PROTECTION					SG-17			
FIRST AID TRAINING				PUB-2017				
FIRST AID TRAINING (FLIP CHART)				PUB-2018				
FLOOR MATS & RUNNERS							ISDS595	
GAS CUTTING EQUIPMENT IN PETROLEUM OPERATION				RP-2009				
GAS DRILLING				RP-2011				
GAS WELDING EQUIPMENT IN PETROLEUM OPERATION				RP-2009				
GEOPHYSICAL EXPLORATION				BULL-2024				
GLASSES	1910.251							
GUARDS, INDUSTRIAL TRUCKS	1910.178							
HATS			Z89.1				ISDS561	
HOT WORK REPAIRS TANK BOTTOM				PSD-2207				
LONGSHORING			MH9.1					
LUMINESCENT & REFLECTING MATERIAL							ISDS245	
METAL & NONMETALLIC OPEN PIT MINES	55.1							
METAL & NONMETALLIC UNDERGROUND MINES	57.1							
MOTION PICTURES							ISDS556A	
NET, CONSTRUCTION							ISDS608	
OCCUPATIONAL, MANAGEMENT POLICY							ISDS585	

SUBJECT	USDL & USDI	NFPA	ANSI	API	MCA	NSC	MISC
	APPLICABLE STANDARDS, GUIDES AND PUBLICATIONS						
(CON'T)SAFETY							
OFF-THE-JOB					SG-11	ISDS601	
OIL REFINERY MAINTENANCE				RP-2007			
PAPER BAG & SACK MANUFACTURE			Z259.1				
PLANT INSPECTION					SG-20		
POLICIES, OCCUPATIONAL						ISDS585	
POSTERS, BULLETIN BOARDS, DISPLAYS						ISDS616	
PULPWOOD LOGGING			O3.1				
RADAR HAZARDS						ISDS481	
RELIEF DEVICES, CARGO & PORTABLE COMPRESSED GAS	1910.167						
RULES, INDUSTRIAL						ISDS535	
SAFETY NET			A10.11				
SAND, GRAVEL, CRUSHED STONE OPERATION	56.1						
SAWMILLS			O2.1				
SEISMIC OPERATION				BULL-2024			
SERVICE STATIONS				APM-5			
SHOES	1910.136		Z41.1				
SLINGS, MANILA ROPE						ISDS259	
SLIPPERY FLOORS						ISDS286	
TROLLEY WIRES - OVERCURRENT PROTECTION	75.1001						
UNDERGROUND COAL MINE	75.1						
WATCHMAN						ISDS212	
WATER BUOYANT DEVICES			Z243.1				UL 1123
WELL PULLING				APM-6			
WOODWORKING MACHINERY			O1.1				
WORKING IN REACTOR				PSD-2211			
SAFETY & HEALTH PROVISIONS	1926.20						
SAFETY BELTS (SEE BELTS, SAFETY &/OR SEAT BELTS)							
LIFELINE REQUIREMENTS	1915.84						
(MULTIPLE REFERENCES)	1916.84						
WORKING MORE THAN 5 FT. ALOFT	1915.47						
(MULTIPLE REFERENCES)	1916.47						
WORKING ON SWINGING SCAFFOLDS	1915.41						
(MULTIPLE REFERENCES)	1916.41						
SAFETY CAN							
METAL - SAFETY STANDARD			Z218.1				UL 30
SAFETY DEVICES							
BASE-MOUNTED DRUM HOISTS	1926.553						
CONVEYORS	1926.555						
HOISTS & ELEVATORS	1926.552		A90.1				
LPG	1926.153						
PNEUMATIC POWER TOOLS	1926.302						
TIRE RACKS	1926.600						

SUBJECT	APPLICABLE STANDARDS, GUIDES AND PUBLICATIONS							
	USDL & USDI	NFPA	ANSI	API	MCA		NSC	MISC
SAFETY FACTOR								
SUSPENSION WIRE ROPES	1926.552							
SAFETY HAT							ISDS561	
SAFETY NETS (SEE NETS, SAFETY)								
SAFETY RELIEF DEVICES								
BULK OXYGEN SYSTEMS	1910.104							
COMPRESSED GAS CYLINDERS	1910.167							
COMPRESSED GASES	1910.107							
FEDERAL SUPPLY CONTRACTS	50-204.71							
FLAMMABLE LIQUID	1910.107							
GASEOUS HYDROGEN SYSTEMS	1910.103							
LIQUEFIED HYDROGEN SYSTEMS	1910.103							
LPG	1910.110							
NON-DOT CONTAINERS	1910.110							
SPRAYING	1910.107							
TANKS, CARGO & PORTABLE	1910.168							
TESTING	1910.167							
(MULTIPLE REFERENCES)	1910.168							
SAFETY STRAPS	1926.959							
FABRICS	1926.959							
SALAMANDERS, SOLID FUEL	1926.154							
SAMPLING								
AIR CONTAMINANTS	71.201							
DUST	70.201							
LIQUID CHEMICALS, TANK CAR					SG-16			
LIQUID CHEMICALS, TANK TRUCK					SG-16			
OIL MIST-PROCEDURE EVALUATION				TB-7201				
PETROLEUM TANKS							ISDS563	
RADIOACTIVE MATERIALS - AIRBORNE			N13.1					
TANK CAR LIQUID CHEMICALS					SG-16			
TANK TRUCK LIQUID CHEMICALS					SG-16			
SAND, GRAVEL, CRUSHED STONE OPERATION	56.1							
AERIAL TRAMWAYS	56.10							
AIR QUALITY	56.5							
COMPRESSED AIR & BOILERS	56.13							
DEFINITION	56.2							
DRILLING	56.7							
ELECTRICITY	56.12							
EXPLOSIVES	56.6							
FIRE PREVENTION & CONTROL	56.4							
GROUND CONTROL	56.3							
ILLUMINATION	56.17							
LOADING, HAULING, DUMPING	56.9							

SUBJECT	USDL & USDI	NFPA	ANSI	API	MCA		NSC	MISC
	APPLICABLE STANDARDS, GUIDES AND PUBLICATIONS							
(CON'T)SAND, GRAVEL, CRUSHED STONE OPERATION								
MAN HOISTING	56.19							
MATERIALS HANDLING & STORAGE	56.16							
PERSONAL PROTECTION	56.15							
ROTARY JET PIERCING	56.8							
SAFETY	56.1							
SAFETY PROGRAMS	56.18							
TRAVELWAYS	56.11							
USE OF EQUIPMENT	56.14							
SAND BLAST NOZZLES	1910.244							
SAND BLASTING	1910.94							
SANDING MACHINES	1910.213							
BARRELS	1910.214							
SANITATION	1926.27							
(MULTIPLE REFERENCES)	1926.51							
(MULTIPLE REFERENCES)	1926.950							
DRESSING ROOMS	1910.141							
EATING OR SMOKING	1915.57							
(MULTIPLE REFERENCES)	1916.57							
(MULTIPLE REFERENCES)	1917.57							
FACILITIES	71.400							
FOOD HANDLING	1910.141							
(MULTIPLE REFERENCES)	1926.51							
GENERAL	1910.141							
HAZARDOUS MATERIALS	1915.57							
(MULTIPLE REFERENCES)	1916.57							
(MULTIPLE REFERENCES)	1917.57							
LUNCH ROOMS	1910.141							
NONWATER CARRIAGE DISPOSAL SYSTEMS			Z4.3					
PERSONAL PROTECTIVE EQUIPMENT	1915.57							
(MULTIPLE REFERENCES)	1916.57							
(MULTIPLE REFERENCES)	1917.57							
PLACES OF EMPLOYMENT			Z4.1					
REST ROOMS	1910.141							
RETIRING ROOMS FOR WOMEN	1910.141							
TEMPORARY LABOR CAMPS			Z4.4					
TEMPORARY SLEEPING QUARTERS	1926.51							
TOILET FACILITIES	1910.141							
(MULTIPLE REFERENCES)	1910.143							
(MULTIPLE REFERENCES)	71.500							
TOILETS	1926.51							
UNCOVERED GARBAGE	1915.57							
WASHING FACILITIES	1910.141							
(MULTIPLE REFERENCES)	1915.57							
(MULTIPLE REFERENCES)	1916.57							
(MULTIPLE REFERENCES)	1917.57							

SUBJECT	USDL & USDI	NFPA	ANSI	API	MCA		NSC	MISC
	APPLICABLE STANDARDS, GUIDES AND PUBLICATIONS							
(CON'T) SANITATION								
WATER	1926.51							
WATER SUPPLY	1910.141							
SAWMILLS								
BINS, BUNKERS, HOPPERS, AND FUEL HOUSES	1910.265							
BLOWER SYSTEMS	1910.265							
BUILDING FACILITIES	1910.265							
BURNERS	1910.265							
CHEMICALS	1910.265							
CHIPPERS	1910.265							
CONVEYORS	1910.265							
DEFINITION	1910.265							
DOCKS	1910.265							
EFFECTIVE DATES	1910.265							
ELECTRIC WIRING	1910.265							
EMERGENCY EXITS	1910.265							
EXHAUST SYSTEM	1910.265							
FIRE ESCAPES	1910.265							
FIRE PROTECTION	1910.265							
FLAMMABLE LIQUID	1910.265							
FLOORS	1910.265							
GAS PIPING AND APPLIANCES	1910.265							
GENERAL REQUIREMENTS	1910.265							
HYDRAULIC SYSTEMS	1910.265							
KILNS, DRY	1910.265							
LADDERS	1910.265							
LIGHTING	1910.265							
LIQUEFIED PETROLEUM GASES	1910.265							
LOG BREAKDOWN	1910.265							
LOG HANDLING, SORTING, AND STORAGE	1910.265							
LUMBER LOADING	1910.265							
LUMBER PILING	1910.265							
LUMBER STORAGE	1910.265	46						
MARKING PHYSICAL HAZARDS	1910.265							
OPEN-SURFACE TANKS	1910.265							
PERSONAL PROTECTIVE EQUIPMENT	1910.265							
PLATFORMS	1910.265							
POWER TRANSMISSION APPARATUS	1910.265							
REFUSE REMOVAL	1910.265							
ROPE, CABLES, SLINGS, AND CHAINS	1910.265							
SANITATION	1910.265							
STACKERS AND UNSTACKERS	1910.265							
STAIRWAYS	1910.265							
STANDARDS SOURCES	1910.265							
(MULTIPLE REFERENCES)	1910.268		02.1					
TANKS	1910.265							
THRESHOLD LIMITS	1910.265							
TRAFFIC CONTROL	1910.265							
TRAMWAYS	1910.265							

APPLICABLE STANDARDS, GUIDES AND PUBLICATIONS

SUBJECT	USDL & USDI	NFPA	ANSI	API	MCA		NSC	MISC
(CON'T)SAWMILLS								
TRESTLES	1910.265							
VATS	1910.265							
VEHICLE	1910.265							
VENTILATION	1910.265							
WALKWAYS	1910.265							
WORK AREAS	1910.265							
SAWS								
BAND	1910.213							
BARREL STAVE	1910.214							
BOLT EQUALIZER, STAVE & HEADING SAWS	1910.214							
CIRCULAR	1910.213							
CROSSCUT SAW, HAND-FED	1910.213							
CUTOFF, SWING SAWS	1910.213							
ELECTRIC HAND							ISDS344	
FORGE FACILITY	1910.218							
HAND, CIRCULAR BLADE TYPE							ISDS344	
HAND, ELECTRIC							ISDS344	
METAL, COLD WORKING							ISDS584	
MILLS	1910.265							
OVERHEAD SWING CUTOFF (& STRAIGHT PULL)							ISDS277	
POWER CHAIN, PORTABLE							ISDS320	
RADIAL	1910.213						ISDS353	
RIPSAW	1910.213							
SWING	1910.213							
TABLE	1910.213							
TILTING ARBOR							ISDS605	
TILTING TABLE							ISDS605	
WOODWORKING BAND							ISDS235	
SCAFFOLDING, WOOD								
CROSS BRACING	1910.28							
DIAGONAL BRACING	1910.28							
FULL DIAGONAL FACE BRACING	1910.28							
GUARDRAILS	1910.28							
LEDGERS	1910.28							
MOVING PLATFORM TO NEXT LEVEL	1910.28							
PLATFORM PLANKS	1910.28							
PLATFORM TURNS (PLANKING)	1910.28							
POLE LOAD REQUIREMENTS	1910.28							
POLE SCAFFOLD 60' OR LESS	1910.28							
POLE SECURE REQUIREMENTS	1910.28							
PUTLOGS OR BEARERS	1910.28							
SPLICED WOOD POLES	1910.28							
WOOD	1910.28							
WOOD POLE	1926.451							

SUBJECT	USDL & USDI	NFPA	ANSI	API	MCA		NSC	MISC
SCAFFOLDS	1915.41							
(MULTIPLE REFERENCES)	1916.41							
(MULTIPLE REFERENCES)	1926.451							
ACCESS	1915.41							
(MULTIPLE REFERENCES)	1916.41							
(MULTIPLE REFERENCES)	1917.41							
BACKRAILS & TOEBOARDS	1915.41							
(MULTIPLE REFERENCES)	1916.41							
(MULTIPLE REFERENCES)	1917.41							
BOATSWAIN'S CHAIR	1910.28							
(MULTIPLE REFERENCES)	1926.451							
BOATSWAIN'S CHAIR, FIBER ROPE	1910.28							
BOATSWAIN'S CHAIR, ROOF IRONS, HOOKS	1910.28							
BOATSWAIN'S CHAIR, SEAT SLINGS	1910.28							
BOATSWAIN'S CHAIR, SIZE REQUIREMENTS	1910.28							
BOATSWAIN'S CHAIR, SPACE REQUIREMENTS	1910.28							
BOATSWAIN'S CHAIR, TACKLE REQUIREMENTS	1910.28							
BOATSWAIN'S CHAIR, WORKMEN PROTECTION	1910.28							
BRICKLAYERS' SQUARE	1910.28							
(MULTIPLE REFERENCES)	1926.451							
BRICKLAYERS' SQUARE, HEIGHT REQUIREMENTS	1910.28							
BRICKLAYERS' SQUARE, LEVEL	1910.28							
BRICKLAYERS' SQUARE, PLATFORM PLANKS	1910.29							
BRICKLAYERS' SQUARE, REINFORCEMENT	1910.28							
BRICKLAYERS' SQUARE, SIZE REQUIREMENTS	1910.28							
BRICKLAYERS' SQUARE, SPACE REQUIREMENTS	1910.28							
BRICKLAYERS' SQUARE, SQUARE REQUIREMENTS	1910.28							
CARPENTERS' BRACKET	1910.28							
(MULTIPLE REFERENCES)	1926.451							
CARPENTERS' BRACKET, BRACKET ATTACHMENT	1910.28							
CARPENTERS' BRACKET, BRACKET DIMENSIONS	1910.28							
CARPENTERS' BRACKET, GUARDRAILS	1910.28							
CARPENTERS' BRACKET, PLATFORM SIZE	1910.28							
CARPENTERS' BRACKET, WORKMEN PROTECTION	1910.28							
CONSTRUCTION	1915.41							
(MULTIPLE REFERENCES)	1916.41							
(MULTIPLE REFERENCES)	1917.41							
CRAWLING BOARDS OR CHICKEN LADDERS	1910.28							
(MULTIPLE REFERENCES)	1926.451							
CRAWLING BOARDS OR CHICKEN LADDERS, CONSTRUCTION	1910.28							
CRAWLING BOARDS OR CHICKEN LADDERS, LIFE-LINE	1910.28							
ELEVATING & ROTATING WORK PLATFORMS	1926.451							
FLOAT OR SHIP	1910.28							
(MULTIPLE REFERENCES)	1926.451							
FLOAT OR SHIP, CONSTRUCTION	1910.28							
FLOAT OR SHIP, PLATFORM DIMENSIONS	1910.28							
FLOAT OR SHIP, UNDER PLATFORM	1910.28							
FLOAT OR SHIP, WOOD DIMENSIONS	1910.28							
FLOAT OR SHIP, WORKMEN	1910.28							
FOLDING LADDER	1910.29							

APPLICABLE STANDARDS, GUIDES AND PUBLICATIONS

SUBJECT	USDL & USDI	NFPA	ANSI	API	MCA		NSC	MISC
(CON'T)SCAFFOLDS								
FOLDING STAIRWAY	1910.29							
FORM	1926.451							
GENERAL REQUIREMENTS	1910.28							
GENERAL REQUIREMENTS, ACCESS LADDER	1910.28							
GENERAL REQUIREMENTS, COMPONENTS	1910.28							
GENERAL REQUIREMENTS, DAMAGED OR WEAKENED	1910.28							
GENERAL REQUIREMENTS, DEBRIS & TRASH	1910.28							
GENERAL REQUIREMENTS, DIAGONAL BRACING	1910.28							
GENERAL REQUIREMENTS, FIBER ROPE	1910.28							
GENERAL REQUIREMENTS, FOOTING OR ANCHORAGE	1910.28							
GENERAL REQUIREMENTS, GUARDRAILS & TOEBOARDS	1910.28							
GENERAL REQUIREMENTS, LOAD CARRYING TIMBER	1910.28							
GENERAL REQUIREMENTS, MATERIALS HOISTING	1910.28							
GENERAL REQUIREMENTS, NAILS & BOLTS	1910.28							
GENERAL REQUIREMENTS, NEEDLE BEAM, IRON WORKERS	1910.28							
GENERAL REQUIREMENTS, OVERHEAD PROTECTION	1910.28							
GENERAL REQUIREMENTS, PLANKING EXTEND REQUIREMNT	1910.28							
GENERAL REQUIREMENTS, PLANKING OVER LPG	1910.28							
GENERAL REQUIREMENTS, PLANKING, WOOD TYPE	1910.28							
GENERAL REQUIREMENTS, POLES, LEGS OR UPRIGHTS	1910.28							
GENERAL REQUIREMENTS, SECURING	1910.28							
GENERAL REQUIREMENTS, SECURING WITH BOLTS	1910.28							
GENERAL REQUIREMENTS, SHORE & LEAN-TO'S	1910.28							
GENERAL REQUIREMENTS, WEATHER WORKING CONDITIONS	1910.28							
GENERAL REQUIREMENTS, WIRE ROPES	1910.28							
HORSE	1910.28							
(MULTIPLE REFERENCES)	1917.41							
(MULTIPLE REFERENCES)	1926.451							
HORSE, CROSS BRACING	1910.28							
HORSE, GUARDRAILS	1910.28							
HORSE, HEIGHT REQUIREMENTS	1910.28							
HORSE, MEMBER DIMENSIONS	1910.28							
HORSE, QUANTITY	1910.28							
HORSE, SPACING	1910.28							
INTERIOR HUNG	1910.28							
(MULTIPLE REFERENCES)	1926.451							
INTERIOR HUNG, GUARDRAILS	1910.28							
INTERIOR HUNG, HANG WOOD	1910.28							
INTERIOR HUNG, OVERHEAD SUPPORTS	1910.28							
INTERIOR HUNG, PLANKING	1910.28							
INTERIOR HUNG, STEEL TUBE & COUPLER MEMBERS	1910.28							
INTERIOR HUNG, STEEL WIRE ROPE	1910.28							
INTERIOR HUNG, SUPPORTED BY WIRE ROPE	1910.28							
INTERIOR HUNG, SUPPORTING BEAMS	1910.28							
LADDER JACK	1910.28							
(MULTIPLE REFERENCES)	1926.451							
LADDER JACK, CONSTRUCTION	1910.28							
LADDER JACK, HEAVY DUTY	1910.28							
LADDER JACK, LIMITED DUTY	1910.28							

SUBJECT	USDL & USDI	NFPA	ANSI	API	MCA		NSC	MISC
(CON'T)SCAFFOLDS								
LADDER JACK, PLATFORM PLANKS	1910.28							
LADDER JACK, WORKMEN, MAXIMUM	1910.28							
LARGE AREA	1926.451							
LOADING	1910.29							
MANUALLY PROPELLED	1926.451		A92.1					
MASONS'	1926.451							
MASONS' ADJ MULT PT SUSP, BEAMS, STRUCT&STRENGTH	1910.28							
MASONS' ADJ MULT PT SUSP, CHANNEL IRON OUTRIGGER	1910.28							
MASONS' ADJ MULT PT SUSP, GUARDRAILS	1910.28							
MASONS' ADJ MULT PT SUSP, HOIST-MACHINES	1910.28							
MASONS' ADJ MULT PT SUSP, INSTALLATION	1910.28							
MASONS' ADJ MULT PT SUSP, LOAD REQUIREMENTS	1910.28							
MASONS' ADJ MULT PT SUSP, MAINT OF BOLTS, FTGS	1910.28							
MASONS' ADJ MULT PT SUSP, OVERHANG	1910.28							
MASONS' ADJ MULT PT SUSP, OVERHEAD PROTECTION	1910.28							
MASONS' ADJ MULT PT SUSP, PLATFORM STRENGTH	1910.28							
MASONS' ADJ MULT PT SUSP, PLATFORM SUPPORTS	1910.28							
MASONS' ADJ MULT PT SUSP, SINGLE OUTRIGGER BEAMS	1910.28							
MASONS' ADJ MULT PT SUSP, STOP BOLTS	1910.28							
MASONS' ADJ MULT PT SUSP, SUSPENSION WIRE ROPES	1910.28							
MASONS' ADJ MULT PT SUSP, WEBS	1910.28							
MASONS' ADJUSTABLE MULTIPLE POINT SUSPENSION	1910.28							
MISCELLANEOUS TYPES	1915.41							
(MULTIPLE REFERENCES)	1916.41							
NEEDLE BEAM	1910.28							
(MULTIPLE REFERENCES)	1926.451							
NEEDLE BEAM, BEAM REQUIREMENTS	1910.28							
NEEDLE BEAM, PLATFORM SPAN	1910.28							
NEEDLE BEAM, ROPE & HANGERS	1910.28							
NEEDLE BEAM, ROPE ATTACHMENTS	1910.28							
NEEDLE BEAM, SUPPORTS	1910.28							
NEEDLE BEAM, TOOLS	1910.28							
NEEDLE BEAM, WORK SURFACES	1910.28							
OUTRIGGER	1910.28							
(MULTIPLE REFERENCES)	1926.451							
OUTRIGGER, ADDITION LEVEL REQUIREMENTS	1910.28							
OUTRIGGER, BEAMS	1910.28							
OUTRIGGER, DRAWING & SPECIFICATIONS ON SITE	1910.28							
OUTRIGGER, FALLING MATERIAL	1910.28							
OUTRIGGER, INBOARD ENDS	1910.28							
OUTRIGGER, PLANKING	1910.28							
PLASTERS, CONSTRUCTION REQUIREMENTS	1910.28							
PLASTERS, DECORATORS & LARGE AREA	1910.28							
PLASTERS, GUARDRAILS	1910.28							
PLASTERS, PLATFORM PLANKS	1910.28							
POLES, METAL	1915.41							
(MULTIPLE REFERENCES)	1916.41							
POLES, WOOD	1915.41							
(MULTIPLE REFERENCES)	1916.41							

SUBJECT	USDL & USDI	NFPA	ANSI	API	MCA		NSC	MISC
	APPLICABLE STANDARDS, GUIDES AND PUBLICATIONS							
(CON'T)SCAFFOLDS								
ROOFING BRACKETS	1910.28							
ROOFING BRACKETS, CATCH PLATFORMS	1910.28							
(MULTIPLE REFERENCES)	1926.451							
ROOFING BRACKETS, CONSTRUCTION	1910.28							
SAFE LOADS	1915.68							
(MULTIPLE REFERENCES)	1916.68							
SAFETY FACTORS	1915.41							
(MULTIPLE REFERENCES)	1916.41							
(MULTIPLE REFERENCES)	1917.41							
SAFETY STANDARD			A10.8					
SINGLE-POINT ADJUSTABLE SUSPENSION	1910.28							
(MULTIPLE REFERENCES)	1926.451							
SINGLE-PT ADJ SUSP, EMERGENCY BRAKES	1910.28							
SINGLE-PT ADJ SUSP, EQUIPMENT	1910.28							
SINGLE-PT ADJ SUSP, GEARS & BRAKES	1910.28							
SINGLE-PT ADJ SUSP, GUARDS, MID-RAIL & TOEBOARD	1910.28							
SINGLE-PT ADJ SUSP, HOISTING MACHINES	1910.28							
SINGLE-PT ADJ SUSP, POWER UNITS	1910.28							
SINGLE-PT ADJ SUSP, SUPPORTING CABLES	1910.28							
SINGLE-PT ADJ SUSP, SUSPENSION METHODS	1910.28							
STONE SETTER ADJ MULT-PT SUSP, BRIDGING SCAFFOLD	1910.28							
STONE SETTER ADJ MULT-PT SUSP, GUARDRAILS	1910.28							
STONE SETTER ADJ MULT-PT SUSP, HOISTING MACHINES	1910.28							
STONE SETTER ADJ MULT-PT SUSP, INSTALLATION	1910.28							
STONE SETTER ADJ MULT-PT SUSP, OUTRIGGERS	1910.28							
STONE SETTER ADJ MULT-PT SUSP, PLATFORMS	1910.28							
STONE SETTER ADJ MULT-PT SUSP, SUPPORTS	1910.28							
STONE SETTER ADJ MULT-PT SUSP, SUSP WIRE ROPES	1910.28							
STONE SETTER ADJ MULT-PT SUSP, SUSPENSION MTRLS	1910.28							
STONE SETTER ADJ MULT-PT SUSP, WORKING LOADS	1910.28							
STONE SETTER ADJUSTABLE MULTIPLE-PNT SUSPENSION	1910.28							
STONECUTTER'S ADJUSTABLE SUSPENSION	1926.451							
SUSPENDED HORSE	1915.41							
(MULTIPLE REFERENCES)	1916.41							
SUSPENDED LADDER	1915.41							
(MULTIPLE REFERENCES)	1916.41							
SWINGING 2-POINT SUSPENSION	1926.451							
TOWER	1910.29							
TUBE & COUPLER	1910.28							
(MULTIPLE REFERENCES)	1926.451							
TUBE & COUPLER (MOBILE), DESIGN	1910.29							
TUBE & COUPLER (MOBILE), ERECTION	1910.29							
TUBE & COUPLER (MOBILE), HANDRAILS	1910.29							
TUBE & COUPLER (MOBILE), LOADING	1910.29							
TUBE & COUPLER (MOBILE), MATERIALS	1910.29							
TUBE & COUPLER, BEARERS	1910.28							
TUBE & COUPLER, CONSTRUCTION REQUIREMENTS	1910.28							
TUBE & COUPLER, HEAVY-DUTY	1910.28							
TUBE & COUPLER, LIGHT-DUTY	1910.28							

SUBJECT	USDL & USDI	NFPA	ANSI	API	MCA		NSC	MISC
(CON'T) SCAFFOLDS								
TUBE & COUPLER, MEDIUM-DUTY	1910.28							
TUBE & COUPLER, RUNNERS	1910.28							
TUBULAR WELDED FRAME	1910.28							
(MULTIPLE REFERENCES)	1926.451							
TUBULAR WELDED FRAME, BRACING	1910.28							
TUBULAR WELDED FRAME, DRAWING & SPEC REQUIREMNTS	1910.28							
TUBULAR WELDED FRAME, ERECTION REQUIREMENTS	1910.28							
TUBULAR WELDED FRAME, FRAMES	1910.28							
TUBULAR WELDED FRAME, FRAMES & ACCESSORIES MAINT	1910.28							
TUBULAR WELDED FRAME, GUARDRAILS	1910.28							
TUBULAR WELDED FRAME, INSPECTIONS	1910.28							
TUBULAR WELDED FRAME, LEG REQUIREMENTS	1910.28							
TUBULAR WELDED FRAME, LOAD REQUIREMENTS	1910.28							
TUBULAR WELDED FRAME, METAL TUBULAR FRAME	1910.28							
TUBULAR WELDED FRAME, PLANKING	1910.28							
TUBULAR WELDED FRAME, SECURING	1910.28							
TUBULAR WELDED FRAME, SPACING OF PANELS	1910.28							
TUBULAR WELDED FRAME, UPLIFT	1910.28							
TWO-POINT SUSPENSION, ACID SOLUTIONS	1910.28							
TWO-POINT SUSPENSION, BEAMS	1910.28							
TWO-POINT SUSPENSION, BLOCKS	1910.28							
TWO-POINT SUSPENSION, GUARDRAILS	1910.28							
TWO-POINT SUSPENSION, HANGERS	1910.28							
TWO-POINT SUSPENSION, HOISTING MACHINES	1910.28							
TWO-POINT SUSPENSION, PLANK-TYPE	1910.28							
TWO-POINT SUSPENSION, ROOF IRONS OR HOOKS	1910.28							
TWO-POINT SUSPENSION, ROPES, SLINGS, HANGERS	1910.28							
TWO-POINT SUSPENSION, SECURING	1910.28							
TWO-POINT SUSPENSION, SIDE STRING	1910.28							
TWO-POINT SUSPENSION, SUSPENSION	1910.28							
TWO-POINT SUSPENSION, SWINGING	1910.28							
WINDOW JACK	1910.28							
(MULTIPLE REFERENCES)	1926.451							
WINDOW JACK, GUARDRAILS	1910.28							
WINDOW JACK, USE	1910.28							
WOOD POLE	1926.451							
SCALES								
SAFETY STANDARD			B157.1					
SCHOOL BUS								
INSPECTION			D7.4					
SCRAP BAILER							ISDS611	

APPLICABLE STANDARDS, GUIDES AND PUBLICATIONS

SUBJECT	USDL & USDI	NFPA	ANSI	API	MCA		NSC	MISC
SCRAPERS							ISDS256A	
SCREENS, SCREENING	1926.351							
(MULTIPLE REFERENCES)	1926.353							
SCREW JACKS	1926.305							
SCUBA DIVING								
INSTRUCTIONS - MINIMUM COURSE			Z86.3					
SEAT BELTS								
EARTHMOVING EQUIPMENT	1926.602							
MOTOR VEHICLE	1926.601							
SEEPAGE PIT	1910.143							
SEISMIC OPERATION								
SAFETY				BULL-2024				
SELENIUM & ITS COMPOUNDS							ISDS578	HG
SERVICES, MEDICAL	1910.151							
SERVICE STATIONS								
COMBUSTIBLE & FLAMMABLE LIQUIDS	1910.106							
DRAINAGE	1910.106							
ELECTRIC WIRING	1910.106							
FIRE PROTECTION	1910.106							
LPG	1910.110							
SAFETY				APM-5				
SEWAGE DISPOSAL, LABOR CAMPS	1910.142							
SEWERS								
ATMOSPHERE							ISDS550	
FLAMMABLE LIQUID		328						
PIPE CLEANING							ISDS577	
SHACKLES & HOOKS	1910.179							
(MULTIPLE REFERENCES)	1915.63							
(MULTIPLE REFERENCES)	1916.63							
(MULTIPLE REFERENCES)	1917.63							
(MULTIPLE REFERENCES)	1926.251							
SAFE WORKING LOADS	1915.68							
(MULTIPLE REFERENCES)	1916.68							
(MULTIPLE REFERENCES)	1917.68							
(MULTIPLE REFERENCES)	1918.65							
USED ALOFT	1918.65							

APPLICABLE STANDARDS, GUIDES AND PUBLICATIONS

SUBJECT	USDL & USDI	NFPA	ANSI	API	MCA		NSC	MISC
SHAFTS & CORES								
HANDLING							ISDS361	
SHAFTING, POWER TRANSMISSION	1910.219							
SHAPER								
METAL							ISDS216	
WOOD							ISDS333	
SHEARS								
ALLIGATOR							ISDS213	
CONSTRUCTION, CARE, USE			B11.4					
SQUARING, METAL							ISDS328	
SHEARING MACHINE								
CLOTH							ISDS609	
SHEET METAL								
HANDLING							ISDS434	
STORAGE							ISDS434	
SHELTERS								
EMERGENCY	75.1500							
LABOR CAMPS	1910.142							
SHIELDING	1926.351							
BATTERY ROOMS & CHARGING	1926.403							
SHIP								
BOILERS	1915.91							
(MULTIPLE REFERENCES)	1916.91							
BUILDING & REPAIRING	1926.30							
COAST GUARD INSPECTED	1918.12							
DECK MACHINERY	1915.94							
(MULTIPLE REFERENCES)	1916.94							
DEFINITION	1918.3							
FIRE SIGNAL		311						
GAS HAZARD CONTROL		306	Z270.1					
PIPING SYSTEMS	1915.92							
(MULTIPLE REFERENCES)	1916.92							
PROPULSION MACHINERY	1915.93							
(MULTIPLE REFERENCES)	1916.93							
PROTECTION DURING CONSTRUCTION		312						
PUBLIC, DEFINITION	1918.3							
PUBLIC, EXEMPTIONS	1918.11							
(MULTIPLE REFERENCES)	1918.12							
SCAFFOLDS	1926.451							
SEA-VALVES			Z254.1					UL 1121
THROUGH-HULL FITTINGS			Z254.1					UL 1121

APPLICABLE STANDARDS, GUIDES AND PUBLICATIONS

SUBJECT	USDL & USDI	NFPA	ANSI	API	MCA		NSC	MISC
SHIP CERTIFICATION								
EXAMINATIONS	1919.30							
HEAT TREATMENT	1919.36							
PROOF TESTS	1919.31							
(MULTIPLE REFERENCES)	1919.33							
(MULTIPLE REFERENCES)	1919.34							
(MULTIPLE REFERENCES)	1919.35							
SAFE WORKING LOADS	1919.29							
SPECIAL GEAR	1919.32							
UNIT PROOF TESTS	1919.27							
(MULTIPLE REFERENCES)	1919.28							
VISUAL INSPECTION	1919.26							
SHIP SCAFFOLDS	1926.451							
SHIPPING								
PETROLEUM PRODUCTS FIRE PROTECTION				RP-2001				
SHOCK								
EFFECTS ON MAN			S3-W-39					
SHOCK SENSITIVE MATERIAL								
MATERIALS HANDLING					SG-7			
STORAGE					SG-7			
SHORE-BASED EQUIPMENT								
ACCREDITATION	1919.50							
(MULTIPLE REFERENCES)	1919.51							
CERTIFICATION	1919.50							
(MULTIPLE REFERENCES)	1919.51							
EXEMPTIONS	1919.50							
NONDESTRUCTIVE EXAMINATIONS	1919.50							
SHORE-BASED MATERIALS HANDLING DEVICES								
ACCREDITED PERSONS, DUTIES	1919.60							
CERTIFICATION	1919.70							
RATINGS	1919.70							
TON	1919.70							
UNIT PROOF TESTS	1919.60							
SHORING, EXCAVATIONS & TRENCHING	1926.650							
CONCRETE & CONCRETE FORMS	1926.700							
(MULTIPLE REFERENCES)	1926.701							
SUPPORTING SYSTEMS	1926.651							
VERTICAL, CONCRETE FORMWORK							ISDS628	

SUBJECT	USDL & USDI	NFPA	ANSI	API	MCA		NSC	MISC
	APPLICABLE STANDARDS, GUIDES AND PUBLICATIONS							
SHREDDERS	1910.261							
SIGNALING, SIGNALS	1926.201							
CRANES & DERRICKS	1926.550							
CRANES & HOISTS	1926.201							
FLAGMEN	1926.201							
SIGNS & TAGS								
ACCIDENT PREVENTION	1910.145							
(MULTIPLE REFERENCES)	1926.200		Z35.1					
(MULTIPLE REFERENCES)			Z35.2					
(MULTIPLE REFERENCES)			Z35.4					
BIOLOGICAL HAZARDS	1910.145							
CAUTION	1910.93							
(MULTIPLE REFERENCES)	1910.145							
CAUTION, RADIOACTIVE MATERIAL	1910.96							
CLASSIFICATION-DANGER, CAUTION, SAFETY	1910.145							
COLOR CODES	1910.144							
CONVEYORS	1926.555							
DANGER	1910.145							
DESIGN & COLOR	1910.145							
DIRECTIONAL	1910.145							
DO NOT START	1910.145							
EFFECTIVE DATES	1910.149							
EGRESS MEANS	1910.37							
EXIT	1910.37							
(MULTIPLE REFERENCES)	1910.145							
EXIT SIGNS	1910.145							
GAS MASK CANISTERS	1910.134							
IN-PLANT SAFETY	1910.145							
IN-PLANT TRAFFIC	1910.145							
INFORMATIONAL	1910.145							
MANLIFT INSTRUCTION SIGNS	1910.68							
OUT OF ORDER	1910.145							
POWERED PLATFORMS	1910.66							
RADIATION WARNING	1910.97							
(MULTIPLE REFERENCES)	1910.145							
RESPIRATORS	1910.134							
SAFETY INSTRUCTION	1910.145							
SLOW-MOVING VEHICLES	1910.145							
SPECIFICATIONS	1910.93							
(MULTIPLE REFERENCES)	1910.145							
STANDARDS SOURCES	1910.147							
SYMBOL	1910.145							
TRAFFIC SIGNS	1910.145							
WORDING	1910.145							

SUBJECT	USDL & USDI	NFPA	ANSI	API	MCA		NSC	MISC
	APPLICABLE STANDARDS, GUIDES AND PUBLICATIONS							
SILICA (AMORPHOUS)								HG
SILICA (FREE)								HG
SILICON								
DIODE							ISDS581	
SINGLE POST SHORES	1926.701							
SITE CLEARING	1926.604							
EXCAVATIONS	1926.651							
FIRE PREVENTION	1926.151							
SKELETON (MECHANO) DECKS	1918.34							
SKIDS							ISDS260A	
SKIFFS, LIFESAVING	1926.106							
SKIN DISEASE								
INDUSTRIAL							ISDS510	
SKYLIGHT FLOOR OPENINGS	1910.23							
SLEEPING QUARTERS, TEMPORARY	1926.51							
SLICERS	1910.263							
SLINGS	1910.264							
(MULTIPLE REFERENCES)	1926.551							
MANILA ROPE, SAFETY							ISDS259	
SAFETY CODE			B30.9					
WIRE ROPE - LOAD							ISDS380A	
SLIPPERY CONDITIONS	1918.91							
SMALL ARMS AMMUNITION	1910.109							
PRIMERS	1910.109							
SMOKELESS PROPELLANTS	1910.109							
STORAGE	1910.109							
SMOG								
PHOTOCHEMICAL				AQM 69-6				
SMOKE								
DETECTOR - PHOTOELECTRIC			A131.4					UL 168
VENTING GUIDE		204						

APPLICABLE STANDARDS, GUIDES AND PUBLICATIONS

SUBJECT	USDL & USDI	NFPA	ANSI	API	MCA		NSC	MISC
SMOKELESS PROPELLANTS	1910.109							
SMOKING	1926.151							
DIP TANKS	1910.108	34						
DUAL COMPONENT COATINGS	1910.107							
EXPLOSIVES	1910.109							
FLAMMABLE LIQUID	1910.106							
INDOOR STORAGE	1910.177							
POWDER COATINGS	1910.107							
PROHIBITION	75.1702							
SPRAYING	1910.107							
STORAGE AREAS	1910.177							
SNATCH BLOCKS								
FAIRLEADS	1918.84							
USE	1918.84							
SNOW REMOVAL, INDUSTRIAL							ISDS402	
SODA ASH (SEE SODIUM CARBONATE)								
SODIUM							ISDS231	
SODIUM, METALLIC					CC-31			
(MULTIPLE REFERENCES)					SD-47			
(MULTIPLE REFERENCES)					TEG			
SODIUM CHLORATE					SD-42			HG
SODIUM CYANIDE					SD-30			
SODIUM DICHROMATE					SD-46			
SODIUM HYDROXIDE (SEE CAUSTIC SODA)								
SODIUM PENTACHLOROPHENOL								HG
SOIL								
EXCAVATIONS	1926.650							
(MULTIPLE REFERENCES)	1926.651							
TRENCHING	1926.650							
(MULTIPLE REFERENCES)	1926.652							
SOLDERING & BRAZING								
HAND							ISDS445A	

SUBJECT	USDL & USDI	NFPA	ANSI	API	MCA		NSC	MISC
	APPLICABLE STANDARDS, GUIDES AND PUBLICATIONS							
SOLDERING LUG								
SAFETY LUG								UL 486
SAFETY STANDARD			C33.5					
SOLIDS, FLAMMABLE & COMBUSTIBLE								
PROPERTIES		325M						
SOLVENT EXTRACTION								
FLAMMABLE LIQUID		36						
SOUND								
AUDIOMETER SPECIFICATIONS			S3.6					IEC 177
BACKGROUND NOISE - AUDIOMETER ROOM			S3.1					
COMPUTATION OF LOUDNESS			S3.4					ISO R357
HEARING LOSS TO NOISE EXPOSURE			Z24-X-2					
MEASUREMENTS - FREQUENCIES & BAND NUMBERS			S1.6					
MEASUREMENTS IN REVERBERATION ROOM			S1.21					
PHYSICAL MEASUREMENT			S1.2					
PNEUMATIC EQUIPMENT			S5.1					
SOUND - PRESSURE LEVEL MEASUREMENT			S1.13					
SOUND LEVEL METERS			S1.4					
SOUND LEVEL METER	1910.95							
SPACE HEATING								
CONSTRUCTION			A10.10					
ELECTRIC			C33.12					UL 573
SPECIAL INDUSTRIES								
AGRICULTURAL OPERATIONS	1910.267							
BAKERIES	1910.263							
COOPERAGE	1910.214							
FORGING	1910.218							
LAUNDRIES	1910.264							
PLASTICS INDUSTRY	1910.216							
PULP, PAPER & PAPERBOARD MILLS	1910.261							
PULPWOOD LOGGING	1910.266							
RUBBER INDUSTRY	1910.216							
SAWMILLS	1910.265							
STANDARDS SOURCES	1910.268							
TEXTILES	1910.219							
(MULTIPLE REFERENCES)	1910.262							
WOODWORKING	1910.213							
SPECIFICATIONS								
ACCIDENT PREVENTION SIGNS & TAGS			Z35.1					
(MULTIPLE REFERENCES)			Z35.2					
(MULTIPLE REFERENCES)			Z35.4					
AUDIOMETER			S3.6					IEC 177
COMMODITY SPECIFICATION FOR AIR			Z86.1					CGA G-7

SUBJECT	USDL & USDI	NFPA	ANSI	API	MCA		NSC	MISC
	APPLICABLE STANDARDS, GUIDES AND PUBLICATIONS							
(CON'T) SPECIFICATIONS								
CONVEYORS	1926.555							
GAMMA-RAY RADIATION DOSIMETER			N13.5					
GAMMA-RAY RADIATION SURVEY INSTRUMENT			N13.4					
GLAZING MATERIAL FOR BUILDINGS			Z97.1					
HOISTS & ELEVATORS	1926.552							
ICC TANK CAR					TC-2			
SCAFFOLDS	1926.451							
STANDARD RAILINGS	1926.500							
TOXICITY OF PAINTS & COATINGS			Z66.1					
WIRE ROPE - MINES			M11.1					
X-RAY RADIATION DOSIMETER			N13.5					
X-RAY RADIATION SURVEY INSTRUMENT			N13.4					
SPECTACLES	1926.102							
SPICE								
DUST EXPLOSION - GRINDING PLANT		656	Z12.9					
SPIDER								
BLACK WIDOW							ISDS258	
SPILL CONTAINMENT	1910.106							
SPLASH BOARDS	1918.94							
SPLICING	1926.955							
SPOUTS, BULK CARGO								
CERTIFICATION	1918.13							
DEFINITION	1918.3							
(MULTIPLE REFERENCES)	1919.2							
EXAMINATIONS	1919.81							
(MULTIPLE REFERENCES)	1919.90							
SPRAY BOOTHS	1910.107							
BAFFLE PLATES	1910.107							
CLEANING	1910.94							
(MULTIPLE REFERENCES)	1910.107							
DISTANCES FROM HAZARDS	1910.107							
SPRINKLERS	1910.107							
SPRAY FINISHING	1910.94							
DESIGN, CONSTRUCTION & VENTILATION			Z9.3					
FLAMMABLE LIQUID		33						

SUBJECT	USDL & USDI	NFPA	ANSI	API	MCA		NSC	MISC
SPRAY FINISHING - FLAMMABLE & COMBUSTIBLE MTRLS	1910.107							
AUTOMOBILE UNDERCOATING IN GARAGES	1910.107							
CLEARANCE	1910.107							
DEFINITION	1910.107							
DISTANCES FROM HAZARDS	1910.107							
DRYING, CURING OR FUSION APPARATUS	1910.107							
ELECTRICAL & OTHER SOURCES OF IGNITION	1910.107							
ELECTROSTATIC HAND SPRAYING EQUIPMENT	1910.107							
FIRE PROTECTION	1910.107							
FIXED ELECTROSTATIC APPARATUS	1910.107							
GROUNDING	1910.107							
OPERATION & MAINTENANCE	1910.107							
ORGANIC PEROXIDES & DUAL COMPONENT COATINGS	1910.107							
POWDER COATINGS	1910.107							
PROTECTION	1910.107							
SAFETY RELIEF DEVICES	1910.107							
SMOKING	1910.107							
SPRAY BOOTHS	1910.107							
STORAGE	1910.107							
VENTILATION	1910.107							
SPRAY PAINTING								
AIRLESS							ISDS548	
ELECTROSTATIC							ISDS468	
SPRINKLER SYSTEMS (SEE FIRE PROTECTION, ETC)								
SPROCKETS & CHAINS	1910.219							
STAIRS, FIXED INDUSTRIAL	1910.24							
CLEARANCE	1910.24							
INDUSTRIAL			A64.1					
LENGTH	1910.24							
MINES, FOR - CONSTRUCTION & MAINTENANCE			M12.1					
OPEN RISERS	1910.24							
REQUIRED	1910.24							
RISE	1910.24							
STRENGTH	1910.24							
TREADS	1910.24							
VERTICAL CLEARANCE	1910.24							
WIDTH	1910.24							
STAIRWAYS	1926.500							
(MULTIPLE REFERENCES)	1926.501							
DEMOLITION	1926.851							

SUBJECT	USDL & USDI	NFPA	ANSI	API	MCA		NSC	MISC
	APPLICABLE STANDARDS, GUIDES AND PUBLICATIONS							
STAIRWAYS, GUARDING	1910.23							
STANDARDS ORGANIZATIONS	1910.40							
STANDARDS SOURCES								
ABRASIVE BLASTING	1910.139							
ACCIDENT PREVENTION SIGNS & TAGS	1910.147							
ACETYLENE	1910.115							
AIR CONTAMINANTS	1910.99							
AIR RECEIVERS	1910.169							
(MULTIPLE REFERENCES)	1910.170							
ANHYDROUS AMMONIA	1910.115							
ASBESTOS	1910.99							
BLASTING AGENTS	1910.115							
BRAZING	1910.253							
CLOTHING, PROTECTIVE	1910.139							
COLOR CODES	1910.147							
(MULTIPLE REFERENCES)	1910.149							
COMBUSTIBLE GAS	1910.115							
COMBUSTIBLE LIQUIDS	1910.115							
COMPRESSED AIR EQUIPMENT	1910.170							
COMPRESSED GAS EQUIPMENT	1910.166							
(MULTIPLE REFERENCES)	1910.170							
COMPRESSED GASES	1910.115							
CRANES	1910.183							
CUTTING	1910.253							
DERRICKS	1910.183							
DIP TANKS	1910.115							
EGRESS MEANS	1910.39							
ENVIRONMENTAL CONTROLS	1910.99							
(MULTIPLE REFERENCES)	1910.147							
EXPLOSIVES	1910.115							
EYE & FACE PROTECTION	1910.139							
FIRE EXTINGUISHERS	1910.157							
(MULTIPLE REFERENCES)	1910.165							
FIRE PROTECTION	1910.165							
FIRE SUPPRESSION EQUIPMENT	1910.165							
FIRST AID	1910.153							
FLAMMABLE LIQUID	1910.115							
FOOT PROTECTION	1910.139							
GAS MASK PROTECTION	1910.139							
GENERAL ENVIRONMENTAL CONTROLS	1910.147							
GUARDING MACHINERY	1910.221							
HAND & PORTABLE POWERED TOOLS & OTHER HAND-HELD	1910.246							
HAZARDOUS MATERIALS	1910.115							
HEAD PROTECTION	1910.139							
HYDROGEN	1910.115							
INDOOR STORAGE	1910.183							
IONIZING RADIATION	1910.99							
LABOR CAMPS	1910.147							

SUBJECT	USDL & USDI	NFPA	ANSI	API	MCA		NSC	MISC
	APPLICABLE STANDARDS, GUIDES AND PUBLICATIONS							
(CON'T)STANDARDS SOURCES								
LADDERS	1910.31							
LIFE SAFETY CODE	1910.39							
LPG	1910.115							
MACHINERY & MACHINE GUARDING	1910.221							
MANLIFTS	1910.69							
MATERIALS HANDLING & STORAGE	1910.183							
MEDICAL	1910.153							
MEDICAL SERVICES	1910.153							
NITROUS OXIDE	1910.115							
NOISE EXPOSURE	1910.99							
NONIONIZING RADIATION	1910.99							
NONWATER DISPOSAL SYSTEMS	1910.147							
OCCUPATIONAL HEALTH & ENVIRONMENTAL CONTROL	1910.99							
OXYGEN	1910.115							
PHYSICAL HAZARDS MARKINGS	1910.147							
PLASTICS INDUSTRY	1910.221							
PLATFORMS, POWERED	1910.69							
POWER TOOLS, HAND & PORTABLE	1910.246							
POWERED INDUSTRIAL TRUCKS	1910.183							
POWERED PLATFORMS, MANLIFTS & VEHICLE-MTD	1910.69							
PROTECTIVE EQUIP	1910.139							
RAILINGS	1910.31							
RESPIRATORY PROTECTION	1910.139							
SAFETY COLOR CODES	1910.147							
SANITATION	1910.147							
SAWMILLS			02.1					
SIGNS & TAGS	1910.147							
SPECIAL INDUSTRIES	1910.268							
SPRAY FINISHING	1910.115							
STANDPIPE & HOSE SYSTEMS	1910.158							
(MULTIPLE REFERENCES)	1910.165							
TANKS, CARGO & PORTABLE	1910.168							
(MULTIPLE REFERENCES)	1910.170							
TOEBOARDS	1910.31							
VEHICLE MOUNTED WORK PLATFORMS	1910.69							
VENTILATION	1910.99							
WALKING-WORKING SURFACES	1910.31							
WALL OPENINGS	1910.31							
WELDING, CUTTING & BRAZING	1910.253							
STANDPIPES & HOSE SYSTEMS	1910.158							
(MULTIPLE REFERENCES)	1926.150							
CONNECTION	1910.158							
DRY IDENTIFICATION	1910.158							
GENERAL REQUIREMENTS	1910.158							
HOSE OUTLETS	1910.158							
HYDROSTATIC TESTS	1910.158							
INSPECTION	1910.158							
INSTALLATION	1910.158							

SUBJECT	USDL & USDI	NFPA	ANSI	API	MCA		NSC	MISC
	APPLICABLE STANDARDS, GUIDES AND PUBLICATIONS							
(CON'T)STANDPIPES & HOSE SYSTEMS								
MARKINGS	1910.158							
PROTECTION	1910.158							
SUPPLY	1910.158							
TESTS & MAINTENANCE	1910.158							
WATER SUPPLY	1910.158							
STATIC CHARGES	1910.219							
(MULTIPLE REFERENCES)	1926.551							
STATIC ELECTRICITY							ISDS547	
OCCUPANCY STANDARDS & PROCESS HAZARDS		77						
PETROLEUM PRODUCTS FIRE PROTECTION				RP-2003				
STEAM VAT (VENEER)	1910.30							
STEEL								
ERECTION	1926.750		A10.13					
HANDLING FOR FABRICATION			Z229.1					
REINFORCING	1926.700							
STRUCTURAL STEEL ASSEMBLY	1926.751							
STEEL PLATE								
HANDLING FOR FABRICATION			Z229.1				ISDS565	
STEPS, BULWARK	1918.21							
STEPLADDERS	1910.25							
(MULTIPLE REFERENCES)	1910.26							
STIBINE								HG
STICKING MACHINES	1910.213							
STODDARD SOLVENT								HG
STOPS, RAILS, OVERHEAD GANTRY CRANES	1910.179							
STORAGE (SEE ALSO MATERIALS HANDLING & STORAGE)								
AMMONIUM NITRATE	1910.109	490						
ANHYDROUS AMMONIA	1910.111		K61.1					CGA G-2.1
ASPHALT SAFETY GUIDE				PSD-2205				
BATTERIES, INDUSTRIAL TRUCKS	1910.178							
BLASTING AGENTS	1910.109							
(MULTIPLE REFERENCES)	1926.904		A10.7					
CARGO & PORTABLE TANK STORAGE	1910.168							
CELLULAR RUBBER		231B						
COMPRESSED GASES	1910.101							
CONTAINER & PORTABLE TANK STORAGE	1910.106							
ELECTRIC STORAGE-TANK WATER HEATER			C33.87					UL 174

SUBJECT	USDL & USDI	NFPA	ANSI	API	MCA		NSC	MISC
		APPLICABLE STANDARDS, GUIDES AND PUBLICATIONS						
(CON'T)STORAGE (SEE ALSO MATERIALS HANDLING & STORAGE)								
ELECTRIC WIRING	1910.106							
EXPLOSIVES	1910.109		A10.7					
FIRE PROTECTION	1910.177							
FIREWORKS		44A						
FLAMMABLE & COMBUSTIBLE LIQUIDS	1910.106							
FLAMMABLE LIQUID	1910.107				SG-3			
FLAMMABLE LIQUID - FARM STORAGE		395						
FLAMMABLE LIQUID, DIP TANK	1910.108	33						
FUR STORAGE - CLEANING		81						
FUSES							ISDS358	
GENERAL - INDOORS		231						
GENERAL - OUTDOORS		231A						
HEAVY OIL SAFETY GUIDE				PSD-2205				
IMPACT SENSITIVE MATERIAL					SG-7			
INDOOR	1910.177							
(MULTIPLE REFERENCES)	1926.151							
(MULTIPLE REFERENCES)	1926.152							
INDOOR, DEFINITION	1910.177							
INDOOR, EFFECTIVE DATES	1910.182							
INDOOR, GENERAL	1910.177							
INDOOR, NO FIRE PROTECTION	1910.177							
INDOOR, ROOMS	1910.106							
INDOOR, STANDARDS SOURCES	1910.183							
LEAKAGE - UNDERGROUND FLAMMABLE LIQUID TANK		329						
LIQUID FUELS	1910.178							
LIQUIDS, CLASSES I, II, & III	1910.106							
LNG		59A	Z225.1					
LPG		58	Z106.1					
LPG CONTAINERS	1910.110							
(MULTIPLE REFERENCES)	1926.153							
LPG DEFINITION	1910.110							
MAGNESIUM		48						
MATERIALS HANDLING	1926.250							
(MULTIPLE REFERENCES)	1926.953							
(MULTIPLE REFERENCES)	1926.957							
MECHANICAL HANDLING EQUIPMENT, INDOOR	1910.177							
OPEN YARD	1926.151							
OUTDOOR	1926.152							
OUTDOOR, FLAMMABLE LIQUIDS	1910.106							
PETROLEUM PRODUCTS FIRE PROTECTION				RP-2001				
PILING, INDOOR	1910.177							
PLASTIC FOAM					SG-5			
PLASTICS		231B						
PYROXYLIN PLASTIC		43						
RACK STORAGE		231C						
RADIOACTIVE MATERIALS	1910.96							
(MULTIPLE REFERENCES)	50-204.28							
REFRIGERATED STORAGE SYSTEMS	1910.111							
SERVICE EQUIPMENT INDOOR BUILDINGS	1910.177							

SUBJECT	USDL & USDI	NFPA	ANSI	API	MCA		NSC	MISC
	APPLICABLE STANDARDS, GUIDES AND PUBLICATIONS							
(CON'T)STORAGE (SEE ALSO MATERIALS HANDLING & STORAGE)								
SHEET METAL							ISDS434	
SHOCK SENSITIVE MATERIAL					SG-7			
SMOKING	1910.177							
SMOKING INDOORS	1910.177							
SULFUR, SOLID							ISDS612	
TANKS	1910.106							
TITANIUM		481						
TORPEDO							ISDS358	
WATER PIPES & FITTINGS							ISDS623	
WOOD CHIPS - OUTDOOR		46A						
STORAGE TANK								
CARGO & PORTABLE TANK STORAGE	1910.168							
CHEMICAL STORAGE TANKS	1910.106							
CLEANING PETROLEUM PRODUCTS				RP-2015				
CONTAINER & PORTABLE TANK STORAGE	1910.106							
CORROSION PROTECTION	1910.106							
DIKES	1910.106							
DRAINAGE	1910.106							
FIRE PROTECTION	1910.106							
GASOLINE - FLAME ARRESTOR			Z222.1					UL 525
GAUGING & SAMPLING PETROLEUM TANKS							ISDS563	
HOT TAPPING (FLAMMABLE LIQUIDS)				PSD-2201				
HOT WORK REPAIRS				PSD-2207				
OIL STORAGE TANKS - FLAME ARRESTOR			Z222.1					UL 525
OPEN-SURFACE, VENTILATION	1910.94		Z9.1					
RECORDS	1910.106							
RUBBER LINED							ISDS492A	
TESTING	1910.106							
TOXIC EFFECT LEADED GASOLINE				PSD-2202				
VENT FLAME ARRESTER				PSD-2210				
VENTING	1910.106							
WELDING (FLAMMABLE LIQUIDS)				PSD-2201				
STOWED CARGO, TEMPORARY LANDING PLATFORM	1918.32							
STRAINER								
RUBBER & PLASTICS							ISDS610	
STRAPPING								
STEEL, FLAT OR ROUND							ISDS315A	
STRETCHERS	1918.96							
STRINGING								
ADJACENT TO ENERGIZED LINES	1926.955							
DEENERGIZED CONDUCTORS	1926.955							

SUBJECT	USDL & USDI	NFPA	ANSI	API	MCA		NSC	MISC
	APPLICABLE STANDARDS, GUIDES AND PUBLICATIONS							
STRONGBACK (SEE BEAMS, HATCH (STRONGBACK))								
STYRENE				TR	CIC-73			
(MULTIPLE REFERENCES)					SD-37		ISDS627	HG
TOXIC EFFECT			Z37.15					
SUBSTATIONS								
ELECTRIC, INDUSTRIAL							ISDS559	
FENCES	1926.957							
SUBSURFACE STRUCTURE								
ATMOSPHERE							ISDS550	
SUCKERS, BULK CARGO								
CERTIFICATION	1918.13							
DEFINITION	1918.3							
(MULTIPLE REFERENCES)	1919.2							
EXAMINATIONS	1919.81							
(MULTIPLE REFERENCES)	1919.90							
SUGAR								
DUST EXPLOSION - PRODUCTION, PACKAGING, HANDLING		62	Z12.6					
SULFUR					CIC-74			
(MULTIPLE REFERENCES)					SD-74			
LIQUIDS, HANDLING							ISDS592	
SOLIDS, HANDLING & STORAGE							ISDS612	
SULFUR CHLORIDES					SD-77			
SULFUR DIOXIDE				TR	CC-78			
(MULTIPLE REFERENCES)					SD-52			
(MULTIPLE REFERENCES)					TEG			HG
CILIARY ACTIVITY OF TRACHEA				MRR-7101				
INJURY - FACTS & OPINIONS				AQM 69-10				
PLANTS				AQM 69-8				
SULFUR FIRE								
DUST EXPLOSION		655	Z12.12					
SULFUR TRIOXIDE					CC-52			
(MULTIPLE REFERENCES)					TEG			HG
SULFURIC ACID				TR	CC-25			
(MULTIPLE REFERENCES)					CC-68			
(MULTIPLE REFERENCES)					CIC-67			
(MULTIPLE REFERENCES)					CIC-75			
(MULTIPLE REFERENCES)					CIC-76			
(MULTIPLE REFERENCES)					SD-20			
(MULTIPLE REFERENCES)					TEG		ISDS325A	HG

APPLICABLE STANDARDS, GUIDES AND PUBLICATIONS

SUBJECT	USDL & USDI	NFPA	ANSI	API	MCA		NSC	MISC
SUPPORTING SYSTEMS	1926.651							
SURFACE								
SURVEYING							ISDS614	
SURFACE BURNING - BUILDING MATERIALS								
FIRE TESTS		255						ASTM E84-68
SURVEYING								
SURFACE							ISDS614	
SUSPENSION SCAFFOLDS	1926.451							
SWEEPING COMPOUNDS								
COMBUSTIBILITY			Z240.1					UL 41
SWITCHES								
CRANES	1910.179							
TRIP & EMERGENCY	1910.216							
SWITCHGEAR								
SAFETY REQUIREMENTS			C37.19					
SYMBOL								
AGRICULTURAL MACHINERY - OPERATOR CONTROLS (MULTIPLE REFERENCES)			B114.2					ASAE S304.2 SAE J389
AGRICULTURAL TRACTOR - OPERATOR CONTROLS (MULTIPLE REFERENCES)			B114.2					ASAE S304.2 SAE J389
VEHICLE IDENTIFICATION - SLOW MOVING (MULTIPLE REFERENCES)			B114.1					ASAE S276.2 SAE J943

APPLICABLE STANDARDS, GUIDES AND PUBLICATIONS

SUBJECT	USDL & USDI	NFPA	ANSI	API	MCA		NSC	MISC
TABLES, TEMPORARY LANDING	1918.32							
TAG LINES	1918.81							
(MULTIPLE REFERENCES)	1926.551							
(MULTIPLE REFERENCES)	1926.955							
TAGS, ACCIDENT PREVENTION	1910.145							
(MULTIPLE REFERENCES)	1926.200							
CONVEYORS	1926.555							
TAGGING OF CIRCUITS	1926.400							
TANKS								
ABOVEGROUND - VENTING	1910.106							
INSIDE - VENTING	1910.106							
PORTABLE - VENTING	1910.106							
UNDERGROUND - VENTING	1910.106							
TANKS, ATMOSPHERIC	1910.106							
TANKS, CARGO								
COMPRESSED GASES	1910.168							
DEFINITION	1910.168							
SAFETY RELIEF DEVICES	1910.168							
TANKS, PORTABLE STORAGE	1910.106							
(MULTIPLE REFERENCES)	1926.152							
COMBUSTIBLES	1910.106							
COMPRESSED GASES	1910.168							
CORROSION PROTECTION	1910.106							
OPEN-SURFACE	1910.94							
TANK CAR								
CAUSTIC POTASH UNLOADING					TC-3			
CAUSTIC SODA UNLOADING					TC-3			
FLAMMABLE LIQUID CLEANING				APM-13				
FLAMMABLE LIQUID UNLOADING					TC-4			
ICC SPECS					TC-2			
LIQUID CHEMICAL SAMPLING					SG-16			
MURIATIC ACID UNLOADING					TC-2			
PHOSPHORIC ACID UNLOADING					TC-2			
PLATFORMS					TC-7			
SAMPLING LIQUID CHEMICALS					SG-16			
TANK TRUCK								
DRIVER'S HANDBOOK				BULL-1609				
LIQUID CHEMICAL SAMPLING					SG-16			
SAMPLING LIQUID CHEMICALS					SG-16			

SUBJECT	USDL & USDI	NFPA	ANSI	API	MCA		NSC	MISC
				APPLICABLE STANDARDS, GUIDES AND PUBLICATIONS				
TANK VEHICLES								
FLAMMABLE LIQUID CLEANING				RP-2013				
FOR FLAMMABLE LIQUIDS		385						
TEFLON, FLUORO RESINS, ETC.								HG
TELLURIUM								HG
TEMPORARY BUILDINGS	1926.151							
TEMPORARY ELECTRIC WIRING								
CONSTRUCTION SITE							ISDS515	
TENONING MACHINES	1910.213							
TENTS		102	Z20.3					
NIGHT	1918.43							
RAIN	1918.71							
TEST EQUIPMENT								
ACCREDITED PERSONS	1919.10							
ELECTRICAL							ISDS496	
TEST SCHEDULE								
ACCREDITED PERSONS	1919.10							
TESTING								
BULK OXYGEN	1910.104							
CRANES	1910.179							
(MULTIPLE REFERENCES)	1910.180							
DERRICKS	1910.18							
ELECTRICAL EQUIP - NUCLEAR POWER GENERATING STA			N45.2.4					IEEE 336
FIRE EXTINGUISHERS, CARBON DIOXIDE	1910.161							
FIRE EXTINGUISHERS, DRY CHEMICAL	1910.160							
FIRE EXTINGUISHERS, PORTABLE	1910.157							
GASEOUS HYDROGEN SYSTEMS	1910.103							
INITIAL	1919.14							
INSTRUMENTATION - NUCLEAR POWER GENERATING STA			N45.2.4					IEEE 336
LIQUEFIED HYDROGEN SYSTEMS	1910.103							
PERIODIC	1919.15							
PIPING	1910.106							
POWERED PLATFORMS	1910.66							
RADIATION ALARM	1910.96							
SAFETY RELIEF VALVES	1910.167							
(MULTIPLE REFERENCES)	1910.168							
SCHEDULE	1919.10							
STORAGE TANK	1910.106							
SUPERVISION	1919.10							

SUBJECT	USDL & USDI	NFPA	ANSI	API	MCA		NSC	MISC
	APPLICABLE STANDARDS, GUIDES AND PUBLICATIONS							
TETRACHLOROETHANE					SD-34			
TETRACHLOROETHYLENE								HG
TOXIC EFFECT			Z37.22					
TETRAETHYLLEAD								HG
TETRAHYDROFURAN								HG
TETRANITROMETHANE								HG
TETRYL							ISDS218	
TEXTILES								
FLAMMABILITY TEST (MULTIPLE REFERENCES)			L14.69					ASTM D1230-61 AATCC 33
PROTECTIVE EQUIP	1910.262							
SAFETY CODE			L1.1					
THERAPY								
INHALATION - GAS SYSTEM		56B						
THORIUM & ITS COMPOUNDS								HG
THRESHOLD LIMIT VALUES (TLV)	1926.55							
TABLE G-2 (TLV)	1910.93							
TICK BITES							ISDS228	
TIMBER, TIMBERING								
EXCAVATIONS	1926.651							
FORMS (CONCRETE)	1926.701							
TRENCHING	1926.652							
TIRE								
MOUNTING, HEAVY DUTY							ISDS411	
SAFETY RACK	1926.600							
TITANIUM								HG
STORAGE & HANDLING		481						
TITANIUM DIOXIDE								HG
TOE BOARDS	1910.23							
SAFETY STANDARD			A12.1					
SCAFFOLDING	1926.451							
STAND SPECIFICATIONS	1926.500							

APPLICABLE STANDARDS, GUIDES AND PUBLICATIONS

SUBJECT	USDL & USDI	NFPA	ANSI	API	MCA		NSC	MISC
TOGGLES, BRIDLE	1918.42							
TOILETS	1926.51							
FACILITIES	71.500							
TOILET ROOMS								
CHEMICAL	1910.143							
COMBUSTION	1910.143							
CONSTRUCTION	1910.141							
FACILITIES	1910.141							
LABOR CAMPS	1910.142							
PORTABLE	1910.143							
RECIRCULATING	1910.143							
TOLUENE				TR	CC-19			
(MULTIPLE REFERENCES)					SD-63			
(MULTIPLE REFERENCES)					TEG		ISDS204	HG
TOXIC EFFECT			Z37.12					
TOLUENE DIISOCYANATE								HG
TOLUIDINE					SD-82			
TOLUOL (SEE TOLUENE)								
TOLYLENE DIISOCYANATE					SD-73			
TOOLS	1926.951							
ABRASIVE BLASTING	1915.23							
(MULTIPLE REFERENCES)	1916.23							
ELECTRIC - SAFETY STANDARD			C33.49					UL 45
ELECTRIC - STATIONARY & FIXED			C33.86					UL 987
EMPLOYEE FURNISHED, HAND & POWER	1910.242							
FASTENING - EXPLOSIVE - ACTUATED			A10.3					
FEDERAL SUPPLY CONTRACTS	50.204.4							
GROUNDING, HAND & POWER	1926.401							
HAND	1926.302							
(MULTIPLE REFERENCES)	1926.951		A10.3					
HAND & POWER	1910.241							
(MULTIPLE REFERENCES)	1926.300							
HYDRAULIC	1926.951							
INSPECTION	1926.951							
LIVE-LINE	1926.951						ISDS498	
MEASURING ROPES	1926.951							
MEASURING TAPES	1926.951							
MECHANICAL PAINT REMOVERS	1915.23							
(MULTIPLE REFERENCES)	1916.23							
PAINTING	1915.24							
(MULTIPLE REFERENCES)	1916.24							
PNEUMATIC	1926.951							

SUBJECT	USDL & USDI	NFPA	ANSI	API	MCA		NSC	MISC
	APPLICABLE STANDARDS, GUIDES AND PUBLICATIONS							
(CON'T) TOOLS								
PORTABLE ELECTRIC	1918.72							
(MULTIPLE REFERENCES)	1926.951							
SWITCHES	1926.951							
TESTS	1926.951							
UNSAFE	1918.72							
TORCHES	1926.350							
TORPEDO								
STORAGE & HANDLING							ISDS358	
TOWER CRANE							ISDS630	
TOXIC EFFECT								
ACETIC ACID			Z37.39					
ACETONE			Z37.32					
BENZENE			Z37.4					
BERYLLIUM & BERYLLIUM COMPOUNDS			Z37.29					
CADMIUM - FUME & DUST			Z37.5					
CARBON DISULFIDE			Z37.3					
CARBON MONOXIDE			Z37.1					
CARBON TETRACHLORIDE			Z37.17					
CHLORINE			Z37.25					
CHROMIC ACID & CHROMATES			Z37.7					
COATINGS & PAINTS			Z66.1					
COLD WORK	1915.12							
(MULTIPLE REFERENCES)	1917.12							
DICHLORODIFLUOROMETHANE			Z37.38					
DICHLOROMETHANE (METHYLENE CHLORIDE)			Z37.23					
DUSTS & GASES			Z37					
ETHYLENE DIBROMIDE			Z37.31					
ETHYLENE DICHLORIDE			Z37.21					
FLUORIDE DUST - INORGANIC			Z37.28					
FLUOROCARBON 11			Z37.36					
FLUOROCARBON 113			Z37.35					
FLUOROCARBON 12			Z37.38					
FORMALDEHYDE			Z37.16					
HOT WORK	1915.13							
(MULTIPLE REFERENCES)	1917.13							
HOUSEKEEPING	1915.51							
(MULTIPLE REFERENCES)	1916.51							
(MULTIPLE REFERENCES)	1917.51							
HYDROGEN FLUORIDE			Z37.28					
HYDROGEN SULFIDE			Z37.2					
LEAD & ITS COMPOUNDS			Z37.11	AQM 69-7				
(MULTIPLE REFERENCES)				MRR-7102				
LEADED GASOLINE STORAGE TANK				PSD-2202				
MANGANESE			Z37.6					
MERCURY			Z37.8					

SUBJECT	USDL & USDI	NFPA	ANSI	API	MCA		NSC	MISC
			APPLICABLE STANDARDS, GUIDES AND PUBLICATIONS					
(CON'T)TOXIC EFFECT								
MERCURY, ORGANO (ALKYL)			Z37.30					
METAL (WELDING, CUTTING & HEATING)	1926.353							
METHANOL			Z37.14					
METHYL BROMIDE (MONOBROMOMETHANE)			Z37.24					
METHYL CHLORIDE			Z37.18					
METHYL CHLOROFORM			Z37.26					
METHYLENE CHLORIDE (DICHLOROMETHANE)			Z37.23					
MONOBROMOMETHANE (METHYL BROMIDE)			Z37.24					
NITROGEN DIOXIDE			Z37.13					
ORGANO (ALKYL) MERCURY			Z7.30					
PAINTS & COATINGS			Z66.1					
PARADICHLOROBENZENE			Z37.27					
PRESERVATIVE COATINGS	1926.354							
STYRENE			Z37.15					
TETRACHLOROETHYLENE			Z37.22					
TOLUENE			Z37.12					
TRICHLORETHYLENE			Z37.19					
TRICHLOROFLUOROMETHANE			Z37.36					
TRICHLOROTRIFLUOROETHANE			Z37.35					
XYLENE			Z37.10					
TRACTORS (SEE ALSO ROPS)	1910.178							
OPERATION & ROLLOVER STRUCTURE							ISDS622	
TRAFFIC CONTROL								
ADJUSTABLE FACE SIGNAL			D10.1					ITE TR1
IN-PLANT TRAFFIC SIGNS	1910.145							
MANUAL			D6.1					
PRE-TIMED, FIXED CYCLE SIGNAL			D11.1					ITE TR2
TRAFFIC ACTUATED CONTROLLER & DETECTOR			D13.1					ITE TR3
TRAILERS	1910.111							
TRAILING CABLES	1926.402							
FLAME RESISTANCE	75.600							
TRAINING & EDUCATION	1926.21							
API FIRST AID INSTRUCTOR TRAINING GUIDE				BULL-2019				
AUTOMOTIVE MECHANIC - BUS/HEAVY TRUCK/TRAILER			D18.2					
AUTOMOTIVE MECHANIC - PASSENGER CAR/LIGHT TRUCKS			D18.1					
COAL MINE - UNDERGROUND	75.160							
FIRST AID				PUB-2017				
FIRST AID (FLIP CHART)				PUB-2018				
MOTION PICTURES FOR SAFETY							ISDS556A	
NONPROJECTED VISUAL AIDS							ISDS564A	
PERSONNEL	1910.96							
(MULTIPLE REFERENCES)	1910.217							
PHOTOGRAPHY FOR THE SAFETY PROFESSIONAL							ISDS619	
POSTERS, BULLETIN BOARDS, DISPLAYS							ISDS616	

SUBJECT	USDL & USDI	NFPA	ANSI	API	MCA		NSC	MISC
	APPLICABLE STANDARDS, GUIDES AND PUBLICATIONS							
(CON'T) TRAINING & EDUCATION								
PROCESS OPERATOR					SG-15			
PROJECTED STILL PICTURES							ISDS574	
SCUBA DIVING - MINIMUM COURSE			Z86.3					
TRAMWAY								
AERIAL - SAFETY REQUIREMENTS			B77.1					
TRANSFORMERS	1926.402							
SPECIALTY - SAFETY STANDARD			C33.4					UL 506
TRANSMITTERS & RECEIVERS								
TOWER, TRANSMITTER	1926.550							
TRANSPORTATION								
AIRCRAFT LOADING WALKWAYS		417						
AIRPORT TERMINAL		416	Z246.1					
AIRPORT WATER SUPPLY		419						
BLASTING AGENTS			A10.7					
EXPLOSIVES	1910.109		A10.7					
FIREWORKS		44A						
HELIPORT CONSTRUCTION & PROTECTION		418						
INDUSTRIAL TRUCKS - POWERED			B56.1					ISO R1074
INDUSTRIAL TRUCKS - POWERED, USE, MAINT, OPERATN		505	B56.2					
LIQUID PETROLEUM PIPING			B31.4					
MOTOR FREIGHT TERMINAL		513						
TRAPDOORS	1910.23							
TREES								
PRUNING, TRIMMING, REPAIRING, REMOVAL			Z133.1					
TRENCH EXCAVATION							ISDS254	
TRENCHING	1926.650							
(MULTIPLE REFERENCES)	1926.652							
TRICHLOROETHANE					SD-90		ISDS456A	HG
TOXIC EFFECT			Z37.26					
TRICHLOROETHYLENE					SD-14		ISDS389	HG
TOXIC EFFECT			Z37.19					
TRICHLOROFLUOROMETHANE								HG
TOXIC EFFECT			Z37.36					

SUBJECT	APPLICABLE STANDARDS, GUIDES AND PUBLICATIONS							
	USDL & USDI	NFPA	ANSI	API	MCA		NSC	MISC
TRICHLOROTRIFLUOROETHANE								HG
TOXIC EFFECT			Z37.35					
TRICRESYL PHOSPHATE								HG
TRIETHANOLAMINE					CIC-77			
TRIETHYLENETETRAMINE					CIC-78			
TRIMMER								
FLATBED, THREE KNIFE							ISDS505	
TRIMMING, PRESSES	1910.218							
TRINITROTOLUENE							ISDS314	HG
TRIPHENYL PHOSPHATE								HG
TRIPS, TWO-HAND	1910.217							
TRIPPING HAZARDS	1918.91							
TROLLEY WIRES								
CUTOUT SWITCHES	75.1000							
OVERCURRENT PROTECTION	75.1001							
TROWELS, POWERED CONCRETE	1926.700							
TRUCKS								
CHANGING & CHARGING BATTERIES	1910.178							
CONSTRUCTION							ISDS330A	
CONVERTED	1910.178							
DERRICKS	1926.952							
DESIGNATION	1910.178							
DIESEL POWERED	1910.178							
DOCKBOARDS, INDUSTRIAL TRUCKS	1910.178							
FIRE PROTECTION	1910.178	512						
FUEL HANDLING & STORAGE	1910.178							
GANTRY							ISDS257	
GASOLINE POWERED	1910.178							
HIGH LIFT, SAFETY GUARDS	1910.178							
INDUSTRIAL	1910.178							
INDUSTRIAL - ELECTRIC BATTERY POWERED			B56.3					UL 583
INDUSTRIAL - INTERNAL COMBUSTION ENGINE			B56.4					UL 558
INDUSTRIAL - POWERED	1910.178							
(MULTIPLE REFERENCES)	1926.602		B56.1					ISO R1074
INDUSTRIAL - POWERED, USE, MAINT, OPERATION		505	B56.2					
LIFT	1918.73							
LIGHTING	1910.178							
LOADING	1910.178							

SUBJECT	USDL & USDI	NFPA	ANSI	API	MCA		NSC	MISC
	APPLICABLE STANDARDS, GUIDES AND PUBLICATIONS							
(CON'T) TRUCKS								
LPG							ISDS479A	
MAINTENANCE	1910.178							
MATERIALS HANDLING & STORAGE	1910.178							
MECHANICAL HANDLING EQUIPMENT	1910.177							
MINING							ISDS330A	
MOTOR FREIGHT TERMINAL		513						
NOXIOUS GASES & FUMES	1910.178							
OFF-HIGHWAY	1926.602							
OPERATION	1910.178							
OPERATOR TRAINING	1910.178							
OUTSIDE STORAGE	1910.178							
QUARRYING							ISDS330A	
RAILROAD CAR	1910.178							
REFUELING	1910.178							
TRAINING OPERATORS	1910.178							
TUBE								
CATHODE-RAY							ISDS240A	
TUBE & COUPLER								
SCAFFOLDS	1926.451							
SHORING	1926.701							
TUBING								
FLEXIBLE - NONMETALLIC			C33.15					UL 3
TUNNELS & SHAFTS	1926.800							
TURNING MACHINES	1910.213							
TURPENTINE							ISDS367	HG

SUBJECT	APPLICABLE STANDARDS, GUIDES AND PUBLICATIONS							
	USDL & USDI	NFPA	ANSI	API	MCA		NSC	MISC
U-GUARDS	1910.219							
UNDERCOATING AUTOMOBILES	1910.107							
UNDERGROUND								
INSTALLATION	1926.651							
LINES	1926.956							
MINE SAFETY	57.1							
UNDERWATER BLASTING	1926.912							
UNGUARDED EDGES & OPENINGS								
DECKS	1915.43							
(MULTIPLE REFERENCES)	1916.43							
HATCH	1915.43							
(MULTIPLE REFERENCES)	1916.43							
LIFEJACKETS	1915.43							
(MULTIPLE REFERENCES)	1916.43							
MANHOLES	1915.43							
(MULTIPLE REFERENCES)	1916.43							
UNIT PROOF TESTS								
ACCESSORY GEAR	1919.28							
(MULTIPLE REFERENCES)	1919.30							
CRANES	1919.28							
(MULTIPLE REFERENCES)	1919.71							
DERRICKS	1919.27							
(MULTIPLE REFERENCES)	1919.73							
EXAMINATIONS	1919.30							
SAFE WORKING LOADS	1919.73							
SHORE-BASED MATERIALS HANDLING DEVICES	1919.60							
WINCHES	1919.27							
UNLOADING								
CAUSTIC POTASH TANK CAR					TC-3			
CAUSTIC SODA TANK CAR					TC-3			
FLAMMABLE LIQUID TANK CAR					TC-4			
MURIATIC ACID TANK CAR					TC-2			
PHOSPHORIC ACID TANK CAR					TC-2			
TANK CAR, CAUSTIC POTASH					TC-3			
TANK CAR, CAUSTIC SODA					TC-3			
TANK CAR, FLAMMABLE LIQUID					TC-4			
TANK CAR, MURIATIC ACID					TC-2			
TANK CAR, PHOSPHORIC ACID					TC-2			
UNSYM DIMETHYL HYDRAZINE (SEE HYDRAZINE/UDMH)								

SUBJECT	USDL & USDI	NFPA	ANSI	API	MCA		NSC	MISC
UPSETTERS	1910.218							
INDUSTRIAL SAFETY DATA SHEET							ISDS466	
URANIUM (NATURAL) AND ITS COMPOUNDS								HG
URANIUM MINE								
RADIATION PROTECTION			N13.8					
(MULTIPLE REFERENCES)			N7.1					
UTILITIES								
ELECTRIC POWER	1915.53							
(MULTIPLE REFERENCES)	1916.53							
(MULTIPLE REFERENCES)	1917.53							
INFRARED HEATING LAMPS	1915.53							
(MULTIPLE REFERENCES)	1916.53							
STEAM SUPPLY & HOSES	1915.53							
(MULTIPLE REFERENCES)	1916.53							
(MULTIPLE REFERENCES)	1917.53							
UTILITY CABLE								
DIRECT BURIED							ISDS607	

SUBJECT	USDL & USDI	NFPA	ANSI	API	MCA		NSC	MISC
		APPLICABLE STANDARDS, GUIDES AND PUBLICATIONS						
VALVES								
AIR RECEIVERS	1910.169							
COMPRESSED GAS CYLINDERS			B57.1					CGA V-1
(MULTIPLE REFERENCES)								ISO R407
ELECTRIC - FOR HAZARDOUS LOCATIONS			C33.83					UL 1002
MAINTENANCE OF HIGH PRESSURE GATE & PLUG							ISDS440	
SEA-VALVES			Z254.1					UL 1121
VALVES, FILLING TANKS (LIQUIDS)	1910.106							
VANADIUM				AQM 69-13				
VANADIUM PENTOXIDE								HG
VAPORS	1910.93							
(MULTIPLE REFERENCES)	1910.106							
(MULTIPLE REFERENCES)	1926.55							
FEDERAL SUPPLY CONTRACTS	50-204.50							
VAPOR DEGREASER							ISDS429	
VAPOR DEGREASING TANKS	1910.94							
VAPORIZERS								
LIQUID HYDROGEN	1910.103							
LIQUID OXYGEN	1910.104							
LPG	1910.110							
VATS, STEAM	1910.30							
VEGETATION								
AIR QUALITY CRITERIA & ATMOSPHERIC FLUORINE				AQM 69-3				
CUTTING & CLEARING							ISDS575	
VEHICLE								
COMMERCIAL, LPG	1910.110							
IDENTIFICATION - SLOW MOVING			B114.1					ASAE S276.2
(MULTIPLE REFERENCES)								SAE J943
MOUNTED WORK PLATFORMS	1910.67							
VENEER								
CLIPPERS							ISDS542	
CUTTERS & WRINGERS	1910.213							
MACHINERY	1910.30							
VENTILATION (SEE ALSO EXHAUST SYSTEMS &/OR FUMES)	1918.93							
(MULTIPLE REFERENCES)	1926.57							
ABRASIVE BLASTING	1910.94		Z9.4					
AIR CONTAMINANTS	1910.93							
ASBESTOS	1910.93							

SUBJECT	USDL & USDI	NFPA	ANSI	API	MCA		NSC	MISC
	APPLICABLE STANDARDS, GUIDES AND PUBLICATIONS							
(CON'T)VENTILATION (SEE ALSO EXHAUST SYSTEMS &/OR FUMES)								
CARBON MONOXIDE	1918.93							
CLEANING OPERATIONS	1915.21							
(MULTIPLE REFERENCES)	1916.21							
COLD WORK	1915.12							
(MULTIPLE REFERENCES)	1917.12							
COOKING EQUIPMENT - COMMERCIAL		96						
DIP TANKS	1910.108	34						
EXHAUST SYSTEM - DESIGN & OPERATION			Z9.2					
EXHAUST, TEST INSTRUMENTS							ISDS431	
FIRE CODE		211	A52.1					
FLAMMABLE LIQUID	1915.25							
(MULTIPLE REFERENCES)	1916.25							
FUMES	1926.55							
GRINDERS	1910.94							
GRINDING, POLISHING, BUFFING OPERATIONS			Z43.1					
GROUNDING	1918.93							
GUARDING BLOWERS	1918.93							
HEAT VENTING GUIDE		204						
LAUNDRIES	1910.264							
MECHANICAL PAINT REMOVERS	1915.23							
(MULTIPLE REFERENCES)	1916.23							
MECHANICAL VENTILATION	75.300							
OPEN-SURFACE TANKS	1910.94		Z9.1					
PAINTING, CONFINED SPACES	1915.24							
(MULTIPLE REFERENCES)	1916.24							
POLISHING	1910.94							
PRESERVATIVE COATINGS	1926.354							
SHIP SPACES	1915.11							
(MULTIPLE REFERENCES)	1917.11							
SMOKE VENTING GUIDE		204						
SPRAY FINISHING	1910.94							
(MULTIPLE REFERENCES)	1910.107		Z9.3					
SPRAY FINISHING AIRFLOW REQUIREMENTS	1910.94							
SPRAY FINISHING MAKE-UP AIR	1910.94							
TEMPORARY HEATING DEVICES	1926.154							
UNDERGROUND MINE	57.5							
VAULT VENTILATOR - SAFETY CODE			B135.1					UL 680
WELDING	1915.31							
(MULTIPLE REFERENCES)	1916.31							
(MULTIPLE REFERENCES)	1917.31							
WELDING & CUTTING	1926.353							
WELDING, CUTTING & BRAZING	1910.252							
VENTING								
ABOVEGROUND TANKS	1910.106							
INSIDE TANKS	1910.106							
PORTABLE TANKS	1910.106							
UNDERGROUND TANK	1910.106							

SUBJECT	USDL & USDI	NFPA	ANSI	API	MCA		NSC	MISC
	APPLICABLE STANDARDS, GUIDES AND PUBLICATIONS							
VENTING EXPLOSIONS	1910.94	68						
VERMIN CONTROL	1910.141							
VERTICAL SLIP FORM	1926.701							
VESSEL								
HOT TAPPING (FLAMMABLE LIQUIDS)				PSD-2201				
WELDING (FLAMMABLE LIQUIDS)				PSD-2201				
VESTS								
BUOYANT	1926.106						ISDS626	
VIBRATION								
EFFECTS ON MAN			S3-W-39					
VINYL ACETATE					CC-21			
(MULTIPLE REFERENCES)					CIC-79			
(MULTIPLE REFERENCES)					SD-75			
(MULTIPLE REFERENCES)					TEG			
VINYLBENZENE (SEE STYRENE)								
VINYL CHLORIDE	1910.93				CC-46			
(MULTIPLE REFERENCES)					CIC-80			
(MULTIPLE REFERENCES)					SD-56			
(MULTIPLE REFERENCES)					TEG			HG
VINYLIDENE CHLORIDE (SEE 1,1-DICHLOROETHYLENE)								
VISIBILITY	1926.551							
VISUAL AID								
MOTION PICTURES FOR SAFETY							ISDS556A	
NONPROJECTED VISUAL AIDS							ISDS564A	
PHOTOGRAPHY FOR THE SAFETY PROFESSIONAL							ISDS619	
POSTERS, BULLETIN BOARDS, DISPLAYS							ISDS616	
PROJECTED STILL PICTURES							ISDS574	

OCCUPATIONAL SAFETY, HEALTH & FIRE INDEX

SUBJECT	USDL & USDI	NFPA	ANSI	API	MCA		NSC	MISC
	APPLICABLE STANDARDS, GUIDES AND PUBLICATIONS							
WALKING-WORKING SURFACES								
BARGE	1918.37							
BARGE OR TOWBOAT	1918.23							
DECKS	1918.43							
DEFINITION	1910.21							
FIXED INDUSTRIAL STAIRS	1910.24							
FIXED LADDERS	1910.27							
GANGWAYS	1918.21							
GENERAL REQUIREMENTS	1910.22							
GUARDING FLOOR & WALL OPENINGS & HOLES	1910.23							
HOUSEKEEPING	1910.22							
MANUALLY PROPELLED LADDER STANDS & SCAFFOLDS	1910.29							
PORTABLE METAL LADDERS	1910.26							
PORTABLE WOOD LADDERS	1910.25							
SAFETY REQUIREMENTS FOR SCAFFOLDING	1910.28							
WALKWAYS, EXCAVATIONS	1926.650							
WALLS								
OPENINGS			A12.1					
OPENINGS, GUARDING	1926.500							
OPENINGS, HOLE PROTECTION, BARRIER	1910.23							
WAREHOUSES								
AMMONIUM NITRATE	1910.109							
FLAMMABLE LIQUID	1910.106							
WARNING DEVICES, OVERHEAD GANTRY	1910.179							
WARNING SIGNS								
CERTIFIED SPACES	1915.15							
COLD WORK	1915.12							
(MULTIPLE REFERENCES)	1917.12							
ELECTRICAL	1915.111							
OBEYING	1915.14							
PROPULSION MACHINERY	1915.93							
WARPERS & BEAMERS							ISDS487A	
WASHING FACILITIES	1910.141							
(MULTIPLE REFERENCES)	1926.51							
HAND	1910.142							
WASTE DISPOSAL								
COMMUNITY DUMPS		602						
EXHAUST MATERIALS	1926.57							
FLAMMABLE LIQUID	1910.106							
HAZARDOUS WASTES					SG-9			
INCINERATORS		82						
NONWATER SYSTEMS	1910.143							

SUBJECT	USDL & USDI	NFPA	ANSI	API	MCA		NSC	MISC
	APPLICABLE STANDARDS, GUIDES AND PUBLICATIONS							
(CON'T)WASTE DISPOSAL								
RADIOACTIVE MATERIALS	1910.96							
(MULTIPLE REFERENCES)	50-204.29							
SAFETY STANDARD			C33.59					UL 430
WASTE MATERIAL	1926.252							
WATCHMAN								
SAFETY							ISDS212	
WATER								
DRINKING WATER	1918.94							
(MULTIPLE REFERENCES)	71.600							
MARINE OPERATIONS	1926.605							
POTABLE	1926.51							
SUPPLY (FOR FIRE PROTECTION)	1926.150							
WORKING OVER OR NEAR	1926.106							
WATER BLASTING								
HIGH PRESSURE							ISDS633	
WATER COOLER								
DRINKING WATER			C33.82					UL 399
REFRIGERATED DRINKING WATER			A112.11.1					ARI 1010
WATER HEATER								
ELECTRIC STORAGE-TANK			C33.87					UL 174
WATER SAFETY								
BUOYANT DEVICES			Z243.1					UL 1123
WATER SUPPLY								
AIRPORT		419						
FIRE PROTECTION	1910.158							
LABOR CAMPS	1910.142							
SANITATION	1910.141							
SPRINKLER SYSTEMS	1910.159							
WATER TANK								
FIRE CODE		22						
WATER TREATING EQUIPMENT								
COOLING TOWERS		214						
WATERPROOFING								
FLOORS		92M						

SUBJECT	USDL & USDI	NFPA	ANSI	API	MCA		NSC	MISC
	APPLICABLE STANDARDS, GUIDES AND PUBLICATIONS							
WEATHER								
POWER PLATFORMS (TYPES F & T)	1910.66							
WEATHERPROOFING	1910.66							
WEBBING, SYNTHETIC (FOR SLINGS)	1926.251							
WEIGHING MACHINES, SYSTEMS								
SAFETY STANDARD			B157.1					
WELDING, CUTTING & BRAZING	1910.252							
APPLICATION, INSTALLATION, & OPERATION	1910.252							
ARC CUTTING/WELDING EQUIPMENT	1910.252							
CARGO GEAR	1919.19							
CONFINED SPACES	1915.31							
(MULTIPLE REFERENCES)	1916.31							
(MULTIPLE REFERENCES)	1917.31							
CONTAINERS FOR COMBUSTIBLES	50-204.72							
CUTTING & BRAZING	1910.251							
EQUIPMENT	1910.252							
EYE PROTECTION	1915.31							
(MULTIPLE REFERENCES)	1916.31							
(MULTIPLE REFERENCES)	1917.81							
(MULTIPLE REFERENCES)	1926.102							
FIRE PREVENTION & PROTECTION	1910.252							
FLAMMABLE LIQUID PIPELINE/VESSEL/TANK				PSD-2201				
FLUORINE COMPOUNDS	1910.252							
GAS SYSTEM		51B						
HEALTH PROTECTION & VENTILATION	1910.252							
HOLLOW METAL STRUCTURES	1915.34							
(MULTIPLE REFERENCES)	1916.34							
(MULTIPLE REFERENCES)	1917.34							
IN AIRCRAFT HANGAR		410E						
INDUSTRIAL APPLICATIONS	1910.252							
INERT-GAS METAL-ARC	1915.31							
(MULTIPLE REFERENCES)	1916.31							
LPG CONTAINERS	1926.153							
METALS, TOXICITY	1915.31							
(MULTIPLE REFERENCES)	1916.31							
(MULTIPLE REFERENCES)	1917.31							
OPERATION	1910.251							
OXYGEN-FUEL GAS SYSTEM	1910.252	51						
PERSONAL PROTECTION	1910.252							
PRESERVATIVE COATINGS	1915.33							
(MULTIPLE REFERENCES)	1916.33							
(MULTIPLE REFERENCES)	1917.33							
PUBLIC PROTECTION	1910.252							
RECORDS	1910.252							
RESISTANCE WELDING EQUIPMENT	1910.252							
RESPIRATORS	1915.31							
(MULTIPLE REFERENCES)	1916.31							
(MULTIPLE REFERENCES)	1917.31							

APPLICABLE STANDARDS, GUIDES AND PUBLICATIONS

APPLICABLE STANDARDS, GUIDES AND PUBLICATIONS

SUBJECT	USDL & USDI	NFPA	ANSI	API	MCA		NSC	MISC
WINDOWS								
CLEANING			A39.1					
CLEANING, POWERED PLATFORM	1910.66							
CRANE CABS	1926.550							
FIRE TESTS		257	A2.6					ASTM E163-65
WINDUPS & LETOFFS								
RUBBER & PLASTICS							ISDS518	
WIRE & CABLE (SEE CABLES & WIRES)								
WIRE CONNECTOR								
SAFETY STANDARD			C33.5					UL 486
WIRE ROPES, WIRE	1910.180							
(MULTIPLE REFERENCES)	1910.265							
(MULTIPLE REFERENCES)	1919.79							
CARGO GEAR	1919.24							
CRANES & DERRICKS	1926.550							
EXAMINATIONS	1919.79							
HOISTS & ELEVATORS	1926.552							
LIMITATIONS	1919.24							
MINES, SPECIFICATIONS & USE			M11.1					
PROHIBITED	1918.52							
PROOF TESTS	1919.33							
RIGGING EQUIPMENT	1926.251							
SAFE WORKING LOADS	1919.33							
SUSPENSION	1926.552							
USE	1918.63							
WIRING								
TEMPORARY	1926.401							
WOMEN'S SANITATION FACILITIES	1910.141							
WOOD								
LOGS - OUTDOOR STORAGE		46B						
LUMBER - OUTDOOR STORAGE		46						
LUMBER YARDS - RETAIL, WHOLESALE		47						
PLANER, POWER FEED							ISDS225	
ROOM							ISDS423	
SHAPER							ISDS333	
TURNING LATHE							ISDS253A	
WOOD CHIPS - OUTDOOR STORAGE		46A						
WOOD FLOUR PLANT								
DUST EXPLOSION		664	Z12.20					

SUBJECT	USDL & USDI	NFPA	ANSI	API	MCA		NSC	MISC
	APPLICABLE STANDARDS, GUIDES AND PUBLICATIONS							
WOODWORKING								
DUST EXPLOSION		664	Z12.20					
WOODWORKING MACHINERY REQUIREMENTS	1910.213							
BAND SAWS & BAND RESAWS	1910.213							
BORING & MORTISING MACHINES	1910.213							
CIRCULAR RESAWS	1910.213							
GROUNDING	1910.243							
HAND-FED CROSSCUT TABLE SAWS	1910.213							
HAND-FED RIPSAWS	1910.213							
INSPECTION & MAINTENANCE OF WOODWORK MACHINERY	1910.213							
JOINTERS	1910.213							
MACHINE CONTROLS & EQUIPMENT	1910.213							
PORTABLE	1910.243							
POWER TOOLS, HAND & PORTABLE	1910.243							
RADIAL SAWS	1910.213							
SAFETY REQUIREMENTS			01.1					
SANDING MACHINES	1910.213							
SELF-FED CIRCULAR SAWS	1910.213							
SWING CUTOFF SAWS	1910.213							
TENONING MACHINES	1910.213							
TOOLS	1926.304							
VENEER CUTTERS & WRINGERS	1910.213							
WOODSHAPERS & SIMILAR EQUIPMENT	1910.213							
WOODWORKING MACHINERY	1910.213							
WORK PLATFORMS								
MOBILE	1910.29							
VEHICLE MOUNTED	1910.67							
WORK PRACTICES, ASBESTOS	1910.93							
WORKING LOAD, DEFINED	1910.29							
WORKING SURFACES								
AISLES	1910.22							
ALOFT	1915.47							
(MULTIPLE REFERENCES)	1916.47							
BOATS	1915.47							
(MULTIPLE REFERENCES)	1916.47							
(MULTIPLE REFERENCES)	1917.47							
COVERS, STEAM VAT	1910.30							
DOCKBOARDS, BRIDGE PLATES	1910.30							
DOCKBOARDS, PORTABLE	1910.30							
DOCKBOARDS, POWERED	1910.30							
FIREBOX FLOORS	1915.47							
(MULTIPLE REFERENCES)	1916.47							
FORGING MACHINES	1910.30							
(MULTIPLE REFERENCES)	1910.218							
HANDHOLDS	1910.30							

APPLICABLE STANDARDS, GUIDES AND PUBLICATIONS

SUBJECT	USDL & USDI	NFPA	ANSI	API	MCA		NSC	MISC
(CON'T)WORKING SURFACES								
LARGE STEAM VATS	1910.30							
RESTRICTED QUARTERS	1915.47							
(MULTIPLE REFERENCES)	1916.47							
VENEER MACHINERY	1910.30							
WRINGERS	1910.213							
(MULTIPLE REFERENCES)	1910.264							

APPLICABLE STANDARDS, GUIDES AND PUBLICATIONS

SUBJECT	USDL & USDI	NFPA	ANSI	API	MCA		NSC	MISC
X-RAYS	1910.96							
DOSIMETER			N13.5					
INDUSTRIAL							ISDS475	
RADIATION SURVEY INSTRUMENT			N13.4					
SAFETY STANDARD - NON-MEDICAL SOURCE			Z54.1					NBS H93
X-RAY DIFFRACTION								
RADIATION SAFETY			N43.2					
X-RAY EQUIPMENT								
SAFETY STANDARD			C33.67					UL 187
X-RAY INSPECTIONS, MECHANICAL PIPING SYSTEMS	1910.252							
XYLENE				TR	CC-45			
(MULTIPLE REFERENCES)					TEG		ISDS204	HG
TOXIC EFFECT			Z37.10					
XYLOL (SEE XYLENE)								

SUBJECT	USDL & USDI	NFPA	ANSI	API	MCA		NSC	MISC
	APPLICABLE STANDARDS, GUIDES AND PUBLICATIONS							
ZINC	1910.252			AQM 69-16			ISDS267	
ZINC OXIDE							ISDS267	HG
ZIRCONIUM PRODUCTION		482M						HG
ZIRCONIUM POWDER					SD-92		ISDS382	